CW01276792

EARTHMOVERS IN SCOTLAND

Mining, Quarries, Roads & Forestry

EARTHMOVERS IN SCOTLAND

Mining, Quarries, Roads & Forestry

David Wylie

Old Pond
PUBLISHING

First published 2016

Copyright © David Wylie 2016

All rights reserved. No part of this publication may be reproduced, stored in a retrieval system, or transmitted, in any form or by any means, electronic, mechanical, photocopying, recording or otherwise, without prior permission of the copyright holder.

Published by
Old Pond Publishing,
An imprint of 5M Publishing Ltd,
Benchmark House,
8 Smithy Wood Drive,
Sheffield, S35 1QN, UK
Tel: +44 (0) 0114 246 4799
www.oldpond.com

A catalogue record for this book is available from the British Library

ISBN 978-1-910456-56-9

Book layout by Servis Filmsetting Ltd, Stockport, Cheshire
Printed by CPI, UK
Photos by David Wylie unless otherwise indicated

Contents

About the Author

David Wylie is a qualified commercial vehicle engineer and is registered with the Engineering Council in the UK. He has had family connections with the construction and plant hire business from a young age.

His interest in photography started in 1987, at the same time Canon launched its EOS system for SLR cameras, and as a Glaswegian he attended part-time classes in photography at Glasgow University.

His first mining images were taken in October 1994 at an opencast fireclay and coal mine at Bathgate. These were of a RH120-C face shovel, Cat D9 dozer and Cat 777C haul truck.

Author standing in the visor of a 26 cu.m bucket fitted to an RH200, one of the largest hydraulic excavators in the UK – at 520 tonnes – at Banks Mining.

His career in photojournalism started in 2009 when he entered a competition run by a weekly construction publication, *Contract Journal* (*CJ*). *CJ*'s online team were looking for the best photograph with an earthmoving theme. Having won the competition, David was quickly given an assignment to visit the biggest opencast coal mine in Scotland – at Scottish Coal's Broken Cross mine just off the M74 motorway – and a two-page report was published.

At this time, images were also being published in the much revered monthly magazine, *Earthmovers*.

David returned to Broken Cross in June 2010 to cover the biggest mining story that year in Scotland for *Earthmovers*, with a massive £23 million order for ten new Caterpillar 136 tonne capacity haul trucks for Scottish Coal. Following this feature and many subsequent assignments, David became a regular contributor to *Earthmovers Magazine*, covering interesting machines, sites and people all over Scotland, as well as the occasional reports on mining machine factories and tradeshows in mainland Europe.

David has received many commissions from the major earthmoving manufacturers to supply images, videos and words for their in-house magazines, advertisements, online media and machine brochures.

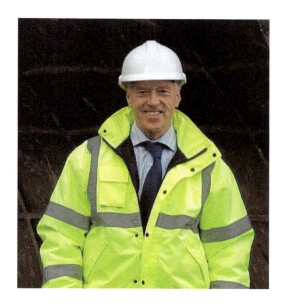

Foreword

Heavy machinery plays a key, but often hidden, part in all of our daily lives. From the roads we drive on to the paper this book is printed on, the supply chain from raw material to finished product involves some of the most technologically advanced equipment on the face of the Earth.

This book provides a snapshot of some of the machinery used across many of Scotland's industries. As such, it provides a unique record of our industrial heritage that, I have no doubt, will be seen by future generations as a historically important document.

I also hope that this publication inspires the current generation of children and young adults to look towards the many aspects of the machinery industry as a worthwhile career. From those who would prefer practical work outdoors to the academically gifted, there are exceptional opportunities; not only for a lifetime of rewarding employment, but also to work with some of the most inspiring and humorous characters that one will ever have the pleasure to meet.

Graham Black, Editor: Earthmovers Magazine.
June 2016

Acknowledgements

I wish to thank many people who provided me with the necessary opportunities, help and support, and who encouraged me to pursue a career in photojournalism.

My interest and development in photography increased rapidly by attending lectures presented by Neil McKellar, a lecturer at Glasgow University. Neil asked us to study – among other things – the old masters of painting in order to understand the use of light and composition to help us capture the subject matter in a powerful image. Thanks also to the rest of the class of 1989–1991 for their friendship, help and advice.

Due to work and family commitments, I gave up photography for a number of years and returned at the start of 2009, with a small digital compact camera that my wife bought me as a present. I soon started visiting various construction sites, mostly beyond the site perimeter, and started uploading my earthmoving images to the photo sharing website Flickr.

That is when a 'sliding door' moment occurred, as Janie Manzoori-Stamford, *Contract Journal* (*CJ*) online content and community editor and Flickr member, invited me to enter a competition in the magazine. I picked a few of my best images and went out to capture something that fitted the theme – earthmoving. I was delighted to win the competition. *CJ* also had a picture page on its website and a photo of the week running in the publication. Janie's colleague, Will Mann, *CJ* web editor, gave me editorial rights to upload my construction images to the magazine's website. Will also gave me my first opportunity in photojournalism, as he arranged a visit and published my first feature, in October 2009, about a large surface coal mine operating 320-tonne Liebherr R9350 excavators at Broken Cross, owned and operated by Scottish Coal.

As well as my activity with *CJ*; Graham Black, editor of *Earthmovers Magazine*, started to publish my images from June 2009 in the readers' section of the monthly title. However, my big break in photojournalism came when Graham and I met at the biannual ScotPlant show in Edinburgh during April 2010, to discuss a possible part-time freelance role for *Earthmovers*. Graham showed a great deal of faith by allowing me to return to Broken Cross to report on one of the biggest stories of the year in Scotland – the delivery and operation of ten new 136-tonne capacity Caterpillar 785D haul trucks; the deal was worth £23million.

I would also like to thank Peter Haddock (of Edson Evers PR), Paul Argent and Keith Haddock for their help, support and advice – and the occasional image – over the years to support some of my features, and to the other *Earthmovers* journalists and the production team that helps to edit and present my work to the readership in the best possible light.

A special thanks to all the owners, operations directors, plant and site managers, press officers, media contacts, drivers and owner drivers, the original equipment manufacturers and their dealers, for their technical information, machine brochures and support. Without everyone's time and commitment, we could not have brought you this book.

Many thanks to Terex's marketing team of Caronne Lockhart and Lyle Sibbald for their extensive trawl through the archives and to the editor of *Blackwood Hodge Memories* website and to Keith Haddock for supplying some of the images and details used in the chapter on the history of the Terex factory at Motherwell, and for the use of Eric C Orlemann's book, *Euclid and Terex Earth-Moving Machines*, as an invaluable research tool. Thanks also to Nigel Rattray for kindly providing some images of Terex machines and of the Demag

The first image published was of my award-winning photo of a CASE CX210 for *Contract Journal*. The image was taken for its competition – best image with an earthmoving theme. The machine was working at Newhouse, just a few miles from my home in Lanarkshire, and the driver kindly helped to pose the falling earth from the bucket.

H485 face shovel. And a special thanks to my daughter, Gillian, who helped with interviews, researched material at the Terex factory and co-authored this particular piece.

Thanks to Gordon Dallas and Stephensons Photographic, Newton Mearns, Glasgow, for the fine work in restoring and making digital images from some of the historic photographic prints found in the mining chapter of this book.

Special thanks to Harry Banks OBE, chairman of the Banks Group, for allowing me to use one of his new mining shovels as the front cover image, and to some of his management team, comprising of Jim Donnelly, Robbie Bentham, Derek Robson, Neil Cook, Ian Ritchie and Darren Banks, for all their support and a warm welcome during multiple visits to their surface mine operations.

Thanks also to Kevin Henderson, Marubeni–Komatsu Ltd, lead field service technician and his team for looking after me over a number of days and a special thank you for supplying some of the images used in the PC2000-8 face shovel assembly chapter.

Thanks to Rachel Turner, commissioning editor, at Old Pond/5M Publishing, who spent hours reading and providing invaluable feedback on my work and also to Alessandro Fratta Pasini, assistant editor; Denise Power, production manager; and Katrina Cacace, marketing and all the other staff at 5M publishing and Old Pond for publishing my book.

Last, but not least, a very special thanks to my wife, Fiona,

First published feature. This image was printed on the front of my two-page report for *Contract Journal* in October 2009 – the massive 320-tonne Liebherr 9350 backhoe at Broken Cross surface mine.

for her unwavering support, as well as proofreading my draft reports, and for reigniting my interest in photography by buying my first digital camera and my SLR film camera back in the late 1980s. And to my two daughters, Gillian and Emma, for their love, and understanding for my interest and passion for all things diggers!

Introduction

From an early age I recall my late father, Sam, working for AV Wilson plant hire in Motherwell as a 360 excavator operator and then for Hewden Stuart Group following a takeover of AV Wilson's business. I also remember him working on his Hymac 580C excavator carrying out general maintenance duties in AV Wilson's yard in Mill Road. That's what operators did then, and some still do now, when off-hire, which wasn't very often in my father's case as he was an excellent operator by all accounts. He could keep the machine on-hire for long periods of time by keeping the site manager – who signed off his weekly hire log – happy by working to a high standard.

He started operating Caterpillar's first 360 machines, in the form of, 215, 225, 235 and 245 excavators, when he moved across to work for Hewden and had worked on some of the biggest projects in Scotland, such as the Megget Reservoir, which is a man-made reservoir in Ettrick Forest, in the Scottish Borders.

Due to my father's work, I was always fascinated with plant and machinery, and was an avid reader of the weekly publication *Construction News*, in particular the section on plant and machinery. I have lasting memories of the large machine adverts that were inserted into the paper during the early 1970s and 1980s. It is fair to say that those big A3 and A2 format fold out glossy posters – from JCB and other manufacturers – still influence some of my photographic captures today and the layout of this book.

As a freelance photojournalist and Scottish correspondent for *Earthmovers Magazine*, I feel privileged to have access to the latest earthmoving machines and technology that are designed to power and control these machines to be the most productive and fuel efficient mobile plant ever made. Some machines featured in this book are record breaking for their time or the first of their kind anywhere in the world.

This book covers numerous site visits across four key applications – mining, quarries, roads and forestry – and I want to share these experiences through not just my own eyes, but also the views of highly skilled and professional people that I have met along the way. I will travel to all four points of the compass across Scotland, to explore the diverse work and visit machines that help to produce energy, everyday products and materials, and support the building and repair of the country's transport network.

For example, a visit to a silica sand quarry reveals the raw materials to make glass that bottles Scotch whisky. And I visit hard stone quarries, where basalt rock is blasted and crushed to protect us from the forces of nature (armoured stone) and to help build sports grounds and provide wear resistant road surfaces. Granite quarries produce red stone for paths, driveways, and high grade rail track ballast for goods and passenger trains to run on. Limestone is extracted to make cement products to build homes, offices and massive road and rail bridges. Surface coal mines help to keep the lights on, as 35% to 40% of our electricity is still generated using this source of fuel. And you may be surprised to know that more than 60% of all the timber produced in the UK is grown and harvested in Scotland.

I have had the pleasure of visiting some of the largest earthmovers in the UK, such as Caterpillar's D11R bulldozer and Liebherr's massive 320-tonne R9350s. On a visit to Banks Mining in West Lothian I was invited to hop a short distance over the Scottish border to see surface mine restoration work and the largest

JCB 110

The only crawler loader of its type. With fully hydrostatic transmission, automatic control system, and all welded integrated chassis, the JCB 110 outdigs and outmanoeuvres conventional machines.

Rear engine design gives 7.85 hp per ton power to weight ratio

Perkins 4.248 four cylinder direct injection engine

Epicyclic final drive

Lifetime lubricated running gear

Superb visibility

One hand control for all track movements

8' 2" dump height

Protected trunnion mounted lift rams

Full welded chassis with integrated track frames

3' 10" forward reach at full height

1¼ cu yd capacity

6 in 1 clamshovel or general purpose bucket

Self levelling bucket linkage

6,500 lb lift capacity to full height

JCB entered the tracked loader market with the 110 in 1972. An example of a large format fold out glossy poster – from JCB and other manufacturers – that influences some of my photographic captures. Photo: JC Bamford Excavators Ltd

hydraulic excavator still working in the UK – the mighty 520-tonne Q&K RH200 at Shotton surface mine in Northumberland. At the other end of the scale, I take a look at a special 1.6-tonne mini digger and its owner operator, as well as manufacturing of mining equipment here in Scotland.

I'm grateful to my hosts, as without their help I would not be able to safely capture some of the close-up action shots found in this book. And I try not to produce just record shots of the machines working, as each one is carefully composed to capture them in their operational setting, and show the power and action of each machine type.

And while there have been a number of high quality books covering earthmoving machines operating in England and Wales, this book sets out to provide a comprehensive look at the Scottish earthmoving scene, with some of the most powerful images presented in large double page format, with incredible

detail to help you get close to the action. I hope you enjoy this book as much as I did gathering the material and writing it.

Much of the preparation for this book was done in the run-up to and aftermath of the 2016 event at Bauma, where one of the star attractions and my assignment for *Earthmovers Magazine*'s July issue, was to cover the 691-tonne Komatsu PC7000-6 face shovel at the show. The exhibition is held in Munich every three years, and attracted 580,000 visitors over seven days. This was up by 60,000 from the last event in 2013 and is a good indicator of how popular this subject now is. I couldn't resist including one or two pictures of the star attraction here. But sadly these pictures were not taken in Scotland or anywhere remotely near it, so although I have many more, I will have to save them for a future book and can only hope a PC7000 will one day find its way to a site here soon or that I see one in action elsewhere in the world!

The largest hydraulic excavator ever to have been displayed at Bauma, Komatsu's new 691-tonne PC7000-6 Super Shovel.

Spacious operator's cab with the driver's seat mounted in the centre, machine performance LCD screen on the L/H side, a 360 degree camera system screen at top right and Modular's ProVision GPS machine guidance system screen bottom right.

Mining

Since 2010 I have been fortunate to visit a number of Scottish surface mines to cover different manufacturers and types of excavators, from Liebherr's massive 320-tonne R9350, O&K RH120-C and E models, Komatsu PC3000s and the forerunner to this machine, the Demag H255S, to smaller machines, such as 20–30-tonne Cat, Volvo and Hyundai coal shovels and dump trucks; the 138-tonne payload Cat 785Ds to 50-tonne payload Bell B50D and large bulldozers from Komatsu and Caterpillar.

I've included two surface mines that I didn't have the opportunity to visit on site, the first chapter covers J Fenton & Son machines working at a fireclay and coal mine at Bathgate in 1994. It was there where I was able to catch my first look, standing at the perimeter, at a massive RH120-C face shovel loading Cat 777 rigid dump trucks and have been fascinated with large mining machines ever since.

I enjoyed working with my daughter, Gillian, as we covered the history and significance of the Terex factory at Motherwell, which has been producing some of the most iconic earthmovers ever made since it opened in 1950.

Scotland has held a few records and operated some of the first machines in their class within the UK's opencast coal sector. I read with interest that Westfield opencast mine near Kinglassie in Fife was reputedly the largest opencast coal mine in the UK with some 26 million tons of coal extracted between 1961 and 1986. It was also claimed to be Europe's deepest mine, at 850ft below ground level at its deepest point. This and more detailed information can be found in Keith Haddock's fantastic book – *British Opencast Coal: A Photographic History 1942–1985*.

The year 1986 was also a significant one for the Scottish opencast coal scene, as the world's largest hydraulic excavator arrived at Coal Contractors Roughcastle coal mine, and with that in mind I was determined to include it and operational details of the site. Information and photographs of this site are scarce; however, I was able to obtain images of it working from Nigel Rattray and Komatsu Mining Germany and detailed knowledge of the site from Derek Taylor, who worked at Roughcastle as an O&K RH9 coal scraper operator.

I was also fortunate to visit at a time when coal prices allowed surface mining companies to make significant investments in new kit, such as Scottish Coal's R9350 excavators and Cat 785D trucks, ATH's five-strong fleet of Komatsu PC3000-6s, and Land Engineering Services 50-tonne Bell B50D trucks used as the site's prime movers. However, towards the end of 2012, some long-established names hit economic headwinds with the slump in world coal prices and high cost of fuel, and they unfortunately went into administration. Scottish Coal and ATH assets were subsequently bought by Hargreaves Services to keep the sites open and carry out restoration work.

That said, Banks Mining bought a new Cat 6030 and Komatsu HD785-7 haul trucks for its Scottish Rusha surface mine in 2012 and at the start of 2016 made a £3.5 million investment in an additional 520-tonne RH200 face shovel, Cat coal trucks and JCB excavators at its Shotton operation.

Once coaling operations come to an end, mining companies are duty bound to restore the site to its former glory or, some cases, leave the land in a much better state than they found it before mining operations started and I have included a number of sites restored to an exceptionally high standard in this chapter.

CHAPTER 1

J Fenton & Son • Northrigg Opencast site Bathgate • October 1994

As my interest in photography took off in the late 1980s I did not have the same access to sites then, as I do now, as a freelance photojournalist and Scottish correspondent for *Earthmovers Magazine*. However, during October 1994 I discovered an opencast site a short drive from my home, next to the A709 road between Armadale and Bathgate – just off Junction 4 of the M8 motorway between Glasgow and Edinburgh – in central Scotland.

Using my first SLR camera, a Canon EOS600, with a roll of Kodak film and fitted with a long focal length 70–210mm lens, I was able to capture my first shots of a huge mining excavator and haul trucks in the shape of two legendary machines of surface mining in the UK – The mighty O&K RH120-C mining excavator and Cat 777C haul truck models. Standing at a low boundary fence, on a grass verge at the side of the A706, I had a clear line of sight into the dig area. I was also fortunate in as much that it must have been early on in the operation, as no perimeter mounds had been built to screen the

Photograph taken standing at the side of the A706 near J4 of the M8 motorway at Bathgate during October 1994 with clear view into the dig area.

The O&K RH120-C has a large operator's cab in keeping with the proportions of the machine and the large front screen provides a good view of the dig area.

site. At that time I was happy with my photographs and left with no details about the mining operation. And in reviewing some of the images used for this piece, I was reminded of how modern full-frame digital cameras, fitted with a professional lens, have transformed our ability to produce high quality images with little or no grain – even in low light – as published in the rest of this book.

Roll forward 22 years; I decided not to sit on these unpublished images any longer and to make them my opening section of this book on mining. However, I now found myself with a challenge of obtaining detailed information about the site. My research for more information lead me to West Lothian Council archive department in Livingston, where I was able to access historical documents, such as planning applications. More than anything else, the large storage box was full of numerous files on environmental impact studies, details of restoration bonds (which is usually a substantial sum of money deposited, as insurance, in case the operator fails to restore the site) and restoration plans.

Initially I thought I would be looking for an opencast coal operator as all I had to go on was the name Fenton displayed on the side of the machines. What I actually found in the files was a brickworks company that had been responsible for the site, and I must congratulate the council staff for being able to locate this archive material with very limited information provided in my request to look at the file they held.

This area can trace its history back to 1859 when John Watson of Glasgow bought the Bathville site in Armadale and shortly after built brickworks to use the local fireclay being mined along with coal. The brickworks changed ownership – and name – many times over the last 100 years, and at the time of my visit it was owned and operated by United Fireclay Products Ltd, who had made the planning application for the Northrigg Opencast site and at the time operated a series of sites in central Scotland.

The 148ha Northrigg Opencast site had been split into a number of different phases and progressively restored.

There were also a number of different contractors hired to work these phases to extract and transport some 250,000 tonnes of clay to its sister brickworks (Armadale Brickworks) nearby using the site's internal haul roads. They extracted around 700,000 tonnes of coal, and this was transported by heavy goods vehicles by road. Records also indicate that the average number of heavy goods vehicles entering or leaving the site was 25 per day. However, due to customer requests by some of the local power generation companies asking for three weeks of coal supply in one day, up to 125 vehicle movements could be required per day with subsequent reduced traffic movements thereafter.

Extraction of both clay and coal also resulted in the moving of some 20 million tonnes of overburden during its eight years of operation – from 1991 to 1999. Records show RJ Budge – which became UK Coal – worked on phase one, which was completed ahead of schedule in the spring of 1994. At the time of my visit in October 1994 it was J Fenton & Son, based in Perth, that was working on the next 42ha phase, referred to as Northrigg Phase W on the planning application documents.

The Armadale and Bathgate area is well known for its coal mining activates, by both surface and underground extraction methods, to reach geological deposits laid down some 500 million years ago. A letter on file stated there had been past underground workings; one seam of Ironstone had been extracted from depths ranging from 220m to 530m until 1955, one seam of fireclay had been mined at a depth of 80–90m, with the last date of extraction being 1919, and seven seams of coal to a depth of 560m being worked until 1967. All these seams were in the vicinity of the site and where you find coal you will usually find clay.

The term 'fireclay' was derived from its ability to resist heat and its original use was in the manufacture of refractories for lining furnaces. However, it is also used in the manufacture of hard-wearing engineering and facing bricks for house building and other construction projects.

As an aside, fireclays are sedimentary mudstones that occur as 'seatearths' underlying almost all coal seams. Seatearths represent the fossil soils on which coal-forming vegetation once grew and are distinguished from associated sediments by the presence of rootlets and the absence of bedding. Fireclays are mainly confined to coal-bearing strata and are commonly named after the overlying coal seam. The production is, therefore, closely related to opencast coal extraction; the seams are typically thin, normally less than 1m, and rarely more than 3metres.

The earthmoving machines I photographed in October 1994 comprised of a 260-tonne RH120-C mining face shovel and a fleet of four Caterpillar 777C haul trucks. The C model truck was launched in the same year, had a payload of 86 tonnes and was powered by a 34.5 litre 870hp V8 turbocharged after-cooled diesel engine. The Cat D9N bulldozer on site was fitted with a single-shank ripper attachment at the rear and an agricultural tractor pulling a fuel bowser. It is a set-up that is not that dissimilar to that operating today in most surface mines, albeit the machines' model numbers may have changed and the kit is far more hi-tech, running clean-burn engines fitted with electronic control systems for both the engine and hydraulics.

In 1994, the RH120-C on site looked like new, but having tracked down some people managing the plant during this time, I discovered this machine was purchased from another operator in Germany. It came with a large 12-tonne steel ball as it had been on quarry operations and was required to extract material and break the oversized rocks by the method called drop-balling (which will be covered in more detail within the quarry chapter – at Dunbar cement works – later in the book). Some of the 777C's haul trucks looked factory fresh, but in fact had been redeployed, from East Chevington opencast coal site in Northumberland, which closed in 1994, to Bathgate. I feel it's fair to say that J Fenton & Sons and its team kept the machines looking and working in tip-top condition.

The bucket sizes on the RH120 series of machines has steadily crept up over the years as engine and hydraulic power has increased and frame design has evolved. The current incarnation is the 300-tonne Caterpillar 6030, is able to handle a 17 cu.m bucket. However, I believe in 1994 this RH120-C would be swinging a 13 cu.m bucket and would be a perfect match to fill a Cat 777C haul truck in four quick passes.

The Cat D9N dozer was observed carrying out three main duties; pushing material over the edge of the overburden tip site, helping to keep the haul road

One Cat 777C is receiving the last pass, as the other 777C is waiting to reverse into the spotting area as soon as possible, in order to keep the big O&K prime mover productive at all times.

Cat D9N on the top of the overburden tip site area.

The first of four passes into the back of the Cat 777C skip.

Fenton kept its kit in good order, however, the single grouser track pads on this D9N look a little worn.

The fuel bowser on its way to refuel the RH120-C prime mover during a planned rest break.

RH120-C has the unique O&K Tri-power system fitted to the boom and connected to the bucket to improve digging and load performance.

in a smooth condition – to save on haul truck tyre wear and fuel burn – and lastly tidying up in and around the loading area of the haul trucks. The reason the loading area needs constant attention from the D9N driver – to remove large rocks – is to minimise damage to the large and expensive tyres.

The planning application for Northrigg Phase W indicated that there were geological problems encountered in Phase one of the site due to glacial action that had disturbed and broken the upper strata, making the extraction of uncontaminated materials difficult. Northrigg Phase W was divided up into three cuts and records show that Cut A, in the south-west section, was the first area to be worked to extract the upper levels of high quality clay and coal. Drill holes indicated that Phase W was situated in reasonably good and level strata, which were dipping north-west at an angle of just four to six degrees and with only minor evidence of glacial disturbance compared to Phase one.

Due to time, cost and efficiency savings, opencast operations, or as they are now referred to as surface mines, are worked by making what is known as a box cut to start the extraction process. Once the topsoils have been removed, the initial box cut can be made and the soils and overburden from it are stored away for reinstatement at the end of operations – this allows the cut void to be progressively restored by backfilling the worked out areas as the box cut moves forward.

Having revisited the exact spot 21 years later, it is hard to believe that the beautifully restored landscape in front of me was once a 40–50m deep hole in the ground with a monster mining shovel, 86-tonne capacity haul trucks and large bulldozers working hard to extract fireclay and coal.

The phased extraction has allowed the development of a sympathetic and regenerated landscape. And possibly without the opencast operation, this may not have changed an otherwise unmanaged, partly industrially degraded landform into a visually acceptable greenspace for future generations. The restoration of surface coal mines is still an important aspect of this industry and will be touched on again in different site visits throughout the mining chapter of this book.

Taken during Oct 2015, 21 years later. It is a nice restoration job, as the RH120-C would have been removing overburden back in 1994, just in front of where a lake has been created to help wildlife to flourish.

CHAPTER 2

Scottish Coal • Caterpillar 785D mining trucks • Broken Cross • June 2010

Scottish Coal was one of the largest surface coal mining companies operating in Scotland and the UK when it took delivery of ten Caterpillar 785D haul trucks at its Broken Cross surface mine. The 785D mining trucks were the first 785s to be sold in the UK for some 15 years. Suffice to say, Finning was extremely pleased at the time to have secured an order for this size of mining truck once again.

On my first assignment representing *Earthmovers Magazine*, I was invited to the official launch event, as the 785Ds trucks were part of a large modernisation programme worth £45 million that started in 2008. Finning (UK) Ltd, Caterpillar's distributor in Great Britain, had supplied some 125 pieces of Cat mining equipment under this programme; the Cat 785Ds deal alone was worth £23 million for the ten mining trucks with a seven-year/36,000 hour repair and maintenance (R&M) equipment support package included.

Broken Cross is situated close to the M74 in South Lanarkshire, Scotland, and the opencast coal mining site covered an area of 610ha, where some 244 million cu.m of overburden was being removed to allow the recovery of 12.8 million tonnes of high quality coal from Scotland's biggest surface mine. The excavation will be 168m at its maximum depth.

The site was mothballed for a time – affected by low coal prices – before recommencing in August 2008. Around 48 million cubic metres of material has been blasted so far at a rate of 400,000 cu.m a week, with some 18,000 tonnes of coal being recovered each week. The new Cat 785Ds are part of a huge earthmoving fleet that is run by Castlebridge Plant, a plant hirer owned by Scottish Resources Group, the parent company of Scottish Coal.

The heavy duty overburden removal is handled by five large 320-tonne Liebherr R9350 hydraulic excavators – the first and only examples found in the UK – which are ideally matched to the new Cat haul trucks with a 136 tonnes payload capability. Scottish Coal invested in the 785Ds to make significant efficiency gains throughout its operations at Broken Cross to achieve the lowest cost per tonne, as well as getting the maximum productivity out of these large hydraulic excavators.

The R9350s are also loading a fleet of more than 40 91-tonne payload haul trucks, such as Cat 777F and Terex TR100, (the Terex trucks are built just 30 miles from the site at the Motherwell factory, further details are covered later in this book) in three passes using their 17 cu.m buckets – three R9350s are operating in backhoe and two in face shovel configuration. During our visit we were invited into the cab of the R9350 face shovel and the operator commented on the ease with which he can load the 785Ds as they present a much larger target area, with a significantly wider dump body skip, in comparison to the 91-tonne class haul trucks, resulting in faster cycle times as the operator can be less precise in placing the excavated material into the truck body.

The R9350 operator was loading the big Cat trucks in five quick passes; however I also observed the R9350 excavator working just that bit harder in trying to turnaround the small queue of mining trucks that was starting to form when a team of 785DS were waiting to be loaded.

Finning had carried out an extensive array of modification to meet Scottish Coal's needs. By listening to and working with the highly experienced management team, a number of important changes were made, such as moving the trucks' front isolator control panel from the right-hand side to the driver side on the bumper

The 785D delivers another load to the tip site. Note the R9350 in background where this truck had just received its load.

The 785D waiting to be loaded. Note the size of the 17 cu.m bucket hitting a smaller target of the Cat 777F's 91-tonne capacity skip.

A Cat 740 articulated dump truck fitted with an adapted coal body hauling the 'black gold' to the coal stockpiling area.

Driver's work station/spacious safety cab. Note the sport car-size steering wheel to operate a massive 250-tonne haul truck. The Caterpillar LCD camera monitor system provides a good view of the rear & offside front wheel station blind spots.

section. The reason for doing so was to improve operational safety; by repositioning the control panel into the driver's view, similarly the fast-fill fuel coupling (567 litres/150 gallons per minute) is placed adjacent to the isolator switch so that when refuelling is carried out the driver has direct view of the operation. In essence, they are trying to engineer out the risk of the truck accidently being driven away, just one of many modifications under a safety first approach.

With safety being the number one priority in any mining operation, Scottish Coal asked Finning to modify the standard mirror package. Again Scottish Coal was very specific about having an effective all-round visibility package on its new trucks, and the Finning team delivered by removing the standard large one-piece 785D mirror units and replacing them with two Cat 777F mirrors, two on each side of the truck. Not only could these be individually angled to cover rear-view issues, but it also meant the gap between the mirrors gave some degree of forward and side visibility. An additional benefit to fitting 777F mirror units meant that they could be heated, which is particularly relevant to working in the Scottish climate!

Operating such a large mining truck as the 785, three close proximity mirrors are also fitted (driver's side, centre, and offside, all mounted to the front guard rails), giving the driver good visibility to the front of the truck. Even with people standing right next to the front bumper they can be easily seen in the mirrors. Finning also supplied and fitted the Caterpillar product camera units, linked to a two monitor system in the cab that covers the offside front wheel station area, rear-view areas; these units are high resolution LCD screens that give the driver a bright clear view of the main blind spots.

At the time, Scottish Coal's MD, Andrew Foster, commented: "The Scottish Coal team have worked closely with Finning to produce a visibility package on the 785D to an industry leading standard. By inspecting the trucks early in the build process we were able to identify simple modifications to make a good truck even better. The safety and comfort of our workforce is our number one priority."

Another bespoke modification was to move the lights out of the front bumper panel area and reinsert them into the engine radiator grille, as UK mining operators tend to park the truck's bumper against a purpose-built mound to allow the drivers to safely step off the trucks during any breaks or shift changes. This is a good illustration of Caterpillar's original design being modified to meet UK market needs by the local Caterpillar dealer. Clearly, if the lights had not been repositioned it would have resulted in impact damage, Scottish Coal also asked Finning to install the fire suppression equipment at bumper level for ease of maintenance.

One of the technical specifications that differs on this class of Caterpillar mining truck from a 777F is the rear axle. Differential and hub reduction gears are fed by a new drive pump system providing continuous filtration and spray lubrication to these components, which means less down time due to longer drain intervals and improved component life. The 785's dual-slope body design was also given a bespoke wear liner package by the Finning team, using Hardox steel plates cut to Finning's own specification templates, to further increase the durability of the truck.

Finning states these trucks are built to be rebuilt, with a planned overhaul at 20,000 hours scheduled and regular oil sampling carried out as part of the periodic maintenance schedule to ensure maximum component life and to detect small problems before they become more expensive ones. In addition to this, Caterpillar's VIMS monitoring system provides critical payload information to the operator and Electronic Technician, or ET in Cat speak (Cat's service tool), data to service personnel at Finning's Glasgow Branch. Finning has a shared financial interest in ensuring maximum uptime for the trucks. While it is early days, the 785s are currently running at 98% availability. Furthermore, they have evidence of Finning's 'lastability' programme working, whereby some B model 777 haul trucks with more than 60,000 hours and having been rebuilt three times were running at 95% availability.

As part of the equipment deal, Finning are also supplying driver training to Scottish Coal's staff, using the instructor seat, which is fitted as standard equipment. Access up to the cab area is via a new step design, over the older 785C model, which is mounted diagonally across the front of the engine radiator resulting in a less steep and safer stepladder arrangement, with cab access into the instructor seat via a separate right-hand side door.

Cat 785D chassis at Finning's Glasgow depot in the early stages of assembly; note the original headlamp position in the bumper.

Dual-sloped mining body, with Hardox liner/wear plates fitted. Finning also repositioned the headlamps from the bumper to the grille area. The truck's isolator control panel and fast-fill fuel coupling were grouped together and moved closer to the driver front window – seen just below the red fire suppression equipment, which was fitted on the bumper top panel for ease of maintenance.

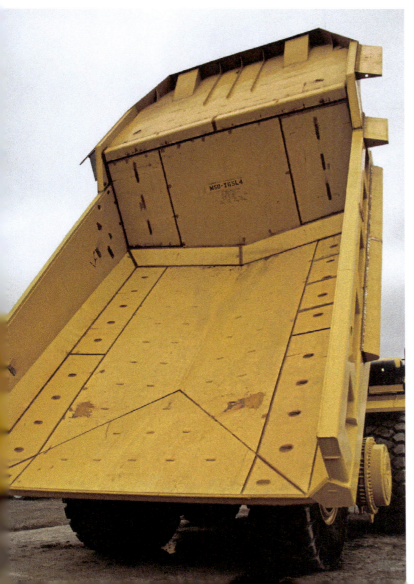

Cat 785D being loaded with overburden by a 320-tonne Liebherr 9350 face shovel.

A Liebherr R9350 loading a Cat 785D haul truck in five passes. Note the tip site in the background.

The instructor seat also provided me with an opportunity to experience first-hand the performance of this large mining truck. The first thing that hits you is the sheer size of the truck, a rise of some 5.67m from the ground to the canopy, operating width of 7.06m, a wheel base of 5.18m and overall length of 11.54m. At this point you can fully understand the reason why an uprated mirror and visibility package was specified. Once settled into the comfy seat, I found the cab to be spacious, and having good all-round visibility with a large glass area and safe ROPS/FOPS five-sided protected work station, with the driver enjoying a comfortable air suspension seat with integral three-point seatbelt. Although the driver's seat has air suspension, the massive size of this truck means it copes extremely well with the haul road that was in less than perfect condition due to adverse weather conditions. The driving and ride experience was one of total comfort, which can be attributed, in part, to the large size and weight, road wheel, and not forgetting the truck's nitrogen-filled suspension units.

Trying to keep the haul road in good condition is hard work with the amount of rainfall Scotland receives. That said, Scottish Coal deploys a number of machines to constantly deal with this issue – a large and highly mobile Cat 834H wheeled-dozer, a Caterpillar grader and a number of Cat D9T dozers all help to keep the haul road surface in tip-top condition.

The 785D is powered by Caterpillar's 3512C HD 12-cylinder, four-stroke engine design that is Tier 2 compliant and which produces a very impressive 1,450hp. Given that the all up gross weight is some 250 tonnes, this power unit provided smooth continuous power throughout the rev range, both in pulling away fully loaded and on the ramps to the tip site areas. During my time in the cab, Caterpillar's six-speed mechanical transmission provided smooth gearshifts with the driver, leaving the gear selector in sixth gear position for forward travel throughout the trip, with the automatic gearbox clearly helping to reduce the driver's workload.

The driver commented that he was delighted to be assigned to a new 785D, having previously operated one of the 91-tonne class haul trucks, and having only operated the truck for a few days he has a very clear opinion that the 785D has superior ride quality, pulling power, low noise levels and a lack of shock loading/movement when being loaded by the big Liebherr's 17 cu.m bucket. It is worth noting that the driver also feels that it's 'his' 785D, as he shares the truck with only one other driver, (during two 12-hour shifts) which helps keeps the cab interior clean and both of them driving it with mechanical sympathy, which makes sense on a vehicle with a price tag of £1.2 million per truck. Regardless of this arrangement, as mentioned previously, Finning and Scottish Coal would soon know if the truck was being driven incorrectly, via a data satellite up-link, and with the information from the Cat VIMS and ET monitoring systems providing both parties with valuable management information.

Finning UK had further supported Scottish Coal's investment in Caterpillar equipment by providing no fewer than 14 Caterpillar-trained engineers dedicated to their operation. These included four engineers working at Broken Cross – on a 12-hour shift basis – with two engineers on day shift and two on night shift duties, to cover the 24-hour mining operation. During our visit there were a number of extra Finning engineers on site putting the finishing touches to the 785s, including mounting the body to the chassis.

The build process was carried out at two main Finning dealership locations, with the dual-sloped specification truck bodies being built in Finning's HQ at Cannock and driven to Broken Cross for final assembly. The 785D chassis was assembled at the Glasgow branch from four individual articulated lorry loads driven from Liverpool docks, having first travelled across 'the Pond' from various plants in America. Some 250 man hours were spent assembling/modifying each truck, which included a final coat of new Caterpillar yellow paintwork at the Glasgow branch. Finning UK invested in a new spray paint booth to accommodate equipment as large as the 785D. The new paint booth, along with four extra service bays and open-plan office space designed to improve inter-departmental communications, was part of a £1.2 million investment by Finning at its Glasgow branch.

Everyone was hoping that this substantial investment would help secure the site's long-term future and I was looking forward to a follow-up visit on the trucks and the site progress. However, with the worldwide slump in coal prices again and higher operating cost

Castlebridge Plant and Finning staff inspecting the latest truck to be completed, which clearly shows the size of this massive mining truck.

– including rising fuel prices – Scottish Coal was put into administration during April 2013. Durham-based Hargreaves Services invested £8.4 million to buy some assets from KPMG, the liquidators of Scottish Coal, to take ownership of the five mines, including Broken Cross in South Lanarkshire, House of Water and Chalmerston in East Ayrshire, St Ninian's in Fife, and Damside in North Lanarkshire. The £8.4 million sale covered much of Scottish Coal's property portfolio, including 30,000 acres of land, together with assets owned by Castlebridge Plant, along with site restoration undertakings, leading to a saving of 500 jobs.

Broken Cross is a vast site and the largest in Scotland producing 1 million tonnes of coal per annum. Look top left to see how small a massive 320-tonne R9350 mining shovel appears in comparison.

A Cat D9T on coal stockpiling duties near the entrance of the mine.

The large, powerful and highly mobile Cat 834H wheeled dozer is tidying up the loading area.

CHAPTER 3

ATH Resources Plc • Komatsu PC3000-6 • Glenmuckloch • March 2011

ATH Resources Plc had invested in five 260-tonne Komatsu PC3000-6 Super Shovels over a five-year period and I went to their Glenmuckloch surface mine operation to enquire how these large German-built hydraulic excavators were coping in the surface coal mines of Scotland.

Glenmuckloch surface mine is near Kirkconnel village, in the Dumfries and Galloway region, and is set back about 1.5 miles from the A76. The site was acquired from Scottish Coal in June 2005, along with the Grievehill site, with an estimated coal reserve of 2.8 million tonnes. The mine commenced operations in the summer of 2006 and started producing coal in September the same year. All coal production from Glenmuckloch is transported to ATH Resources' Crowbandsgate railhead via a 12.2km overland conveyor, known as The Lochside Runner, which is the longest of its type in Europe and can transport up to 500 tonnes of coal per hour.

At the start of excavations in 2006, ATH Resources

The Lochside Runner overland conveyor transported coal to Crowbandsgate railhead. It is a distance of 12.2km, which is the longest conveyor of its type in Europe and can transport up to 500 tonnes of coal per hour.

One of six new Cat D9T dozers purchased by ATH. During the visit I spotted a pre-production Cat 777G model on field trial, which Caterpillar calls a 'field follow' machine. Note the strata starting to go vertical on the top benches.

deployed new PC3000-6 Super Shovels and an O&K RH120C and two Terex/O&K RH120Es, the O&K RH series machine being redeployed to the new Netherton surface mine in Cumnock, East Ayrshire, at the start of 2011 – a site that will be covered in full later on in this chapter.

The three Komatsu PC3000-six models operated at Glenmuckloch can trace their DNA back to the Demag H255S, the prototype H255S being operated at ATH's Skares surface mine and still going strong with some 40,000 hours on the clock. The main production of Komatsu large mining shovels, from the PC3000 to the mighty 720-tonne PC8000, are built by Komatsu Mining Germany GmbH (KMG) at the former Demag factory in Düsseldorf, Germany. ATH's machines were sold and are supported through UK-based KMG Warrington.

Part of our visit was timed to coincide with ATH's planned maintenance tasks during Saturday afternoon when the sunshine miners finished for the day. The on-site maintenance team at Glenmuckloch are highly experienced and with a wealth of knowledge

of these large mining machines. They refer to the Demags affectionately and speak highly of these monster shovels.

Stuart Bickley, maintenance foreman on site, has nothing but praise for the PC3000s due to their ultra-reliable performance and said: "These excavators can be very reliable and can go days without putting a spanner on them, except for routine maintenance, and are achieving 99% availability. However, the same cannot be said for some competitor machines, which in our experience are higher maintenance."

Stuart comments further: "I have one fitter to look after the three PC3000s, but would require a fitter for each of the other non-Komatsu mining shovels to keep them serviceable."

Doug Hogarth (chargehand) is Stuart's right-hand man and echoes Stuart's feelings towards the three Komatsu PC3000s. Doug said: "We have enjoyed really good up-time with these machines over the past five years, with no major issues and there have been no cracks on the stick, boom, or undercarriage." Doug was

quick to point out: "Even some of the main hydraulic hoses on the five-year-old (Fleet No. 3) machine, with 24,000 hours on the clock have not been replaced."

John Taylor (lead fitter on the Komatsus) also went on to explain the PC3000s have a nice simple design to the majority of components. Some examples are: all the electrical systems on a PC3000 are either a 12v or 24v supply, whereas an RH120's hydraulic control system with electrical solenoids is measured in milliamps, and in comparison, the Komatsu has a pilot-controlled hydraulics system, thereby making the PC3000 a much easier machine to maintain and carry out fault diagnostics on.

To help with regular maintenance tasks and to keep the external moving parts in good order, Komatsu has designed the PC3000-6 with an autolube grease system that is supplied from a refillable 200 litre barrel. A second, identical system supplies open gear lubricant through a lube pinion to the very expensive slew ring teeth. Replenishment of the barrels is through the

drop down service centre under the rear of the upper structure, along with all other fluid changes (Wiggins connectors) being mounted here too, including the frequently used refuelling point.

It is impressive to see by how dry and oil leak-free are all three PC3000 machines. This can be put down, in part, to regular maintenance and cleaning, but also the efficient hydraulic oil cooling packs, which would appear to be performing well in keeping the oil temperature under control (only 70°C showing on the oil temperature gauge) and thereby not overstressing the hydraulic oil, connection 'O' rings and hoses.

Some of the regular maintenance tasks consist of hydraulic oil filter changes at 1,000 hour intervals and engine oil change at 250 hours (4,000 hours oil changes are an option with the Centinel system operating), The only major work that has been carried out to date, as with most mining excavators, has been to the PC3000-6 front end equipment. In one example, this involved removing the ESCO lip and teeth on the PC3000

An appointment with the 'dentist' during weekend servicing. Note the worn out bucket teeth lip shrouds laying on the floor of the workshop.

Looking from the top of the overburden pile in to a deep void and the top benches where a PC3000 backhoe is being maintained during planned down time.

backhoe bucket and replacing these with a 140mm thick Komatsu /Hensley lip and wear package.

KMG provides its customers with very good aftersales support from its Warrington HQ. A total of £85,000 worth of KMG parts held on site as imprest stock. In addition to this, ATH has access to imprest parts held at other local surface mining companies and if all else fails, KMG has been known to drive urgent parts direct from the German factory to Scotland within 24 hours. On-site technical training and assistance is also provided by KMG Warrington staff. The maintenance team at Glenmuckloch has also attended a number of maintenance courses at the Komatsu Super Shovel training centre at the Düsseldorf factory.

David Lancashire, group plant director, says of the performance of the PC3000-6: "It would be fair to say that, to date, the PC3000s have exceeded our expectations regarding their availability performance and the cost of maintenance. And although there have been some unscheduled or premature failures, during the five years that we have operated the fleet, the highly effective support we receive from the relatively small team at KMG Warrington has always minimised the effects."

The PC3000-6 backhoe is fitted with an 8.53m long boom and 3.96m stick. The face shovel has a 5.79m boom and 4.26m stick, and both machines are fitted with a 15 cu.m bucket that is powered by a single Komatsu SSA12V159 direct injection, quad turbocharged and after-cooled, 12-cylinder diesel engine, producing an impressive 1,260hp at 1800rpm. Bolted on to the back of this large engine is a PTO gearbox which drives three identical main hydraulic pumps which draw oil from an unpressurised hydraulic tank. Open circuit hydraulics provides maximum cooling and filtering efficiency. The main hydraulic pumps are capable of pumping 3 × 910 litre/min at a maximum relief valve pressure of 310 bar, or 4,495psi in old money. While the PC3000-6 has been fitted with a fire suppression system, Komatsu has designed a firewall between the hydraulic pumps and the engine bay to further reduce the risk of a major fire occurring, i.e. putting a firewall between hot and flammable hydraulic oil and a very hot set of turbochargers and exhaust manifolds. The enormous

4,500 litre fuel tank is topped up at the start of each 12-hour shift, with approximately 1,900 litres of diesel. We were advised the fuel consumption is good at 170–180 litres per hour and is more economical than some twin engine machines.

Komatsu provides the operator with a large comfortable Falling Object Protection System (FOPS) Level 2 cab. The structure is vibration-isolated with 15 viscous mounting pads, which means during hard digging the PC3000 cab moves in a well-damped motion to help make it a comfortable operating environment for the driver. Komatsu claims the cab has a good sound insulation performance at 75dB(A), which is on par with the very latest Cat 777F dump trucks on site.

The cab is equipped with automatic climate control and is pressurised to help with dust out. The operator enjoys a high specification multi-adjustable air suspension seat, redesigned for mining operations and electrically heated, and has a lap seatbelt for safety. There is also a trainer seat with lap-belt. Low effort joystick controls the hydraulic front end equipment, with foot pedals to control the face shovel bucket visor, (bucket clam) travel motors and swing brake.

Full instrumentation and Electronic Text Monitoring (ETM) are provided. An AM/FM radio with CD player is included. The PC3000-6 has a large front windscreen with glass from floor to ceiling and has twin, two-speed wipers to keep the screen clear. The two externally mounted and heated rear-view mirrors work with the external metal sun louvres, however, ATH has removed the left-hand metal louvres to help the driver clean the side windows. All windows are tinted Parsol green and some have internal pull-down sun blinds to help with glare during spells of beautiful Scottish sunny weather, as experienced during our visit – rare but true!

At the end of the morning rest break, I had the rare opportunity to find out first-hand how the PC3000 backhoe performs from the comfort of the instructor seat. Access is via the hydraulic drop down ladder on the upper structure and fixed steps to gain access to the über-size cab.

Once we were both buckled up, the operator tracked the machine up on to the newly blasted material and set to work loading the 91-tonne capacity Cat 777F haul trucks in four quick passes, depending on material density. Bucket fill ranges between 20 to 30 tonnes, as

The PC3000s are refuelled at least once per day using the drop down fast fill services connection below the upper structure.

Large square upper structure with twin cooling packs up front, with the hydraulic pumps connected to the Komatsu engine at the rear.

Hydraulic oil filter changes are required at 1,000-hour intervals.

The PC3000 has a comfortable and spacious cab with a trainer seat behind the driver. The sheer size of the PC3000 and the large front screen affords a good view into the skip. This is despite the PC3000 operating from a lower than normal bench height.

indicated on the haul trucks' load monitoring system. The driver gave good feedback on the level of comfort the Komatsu cab provides, such as low vibration, low noise levels and the effective air con system.

The stability of the PC3000-6 comes in for praise too, as the backhoe version is fitted with the optional 1 metre wide cast – and heat hardened – steel track pads, whereas the face shovel machine is shod with standard 800mm wide track pads. The driver, like the maintenance team, thought the big Komatsu has great reliability. However, having driven a competitor, twin engine machine, he felt that had a performance advantage in digging, but was less reliable and, on balance, the PC3000-6 is a good tool for the job in hand.

ATH Resources gained a RoSPA Gold award during 2010 for safety and with a safety first approach – constantly looking for safety improvements – ATH decided to add to the visibility package of its fleet of PC3000-6, by fitting two large TFT monitors, supplied by Spillards, to cover off the rear and offside blind spots of the machine.

At the time of my visit, the 270ha Glenmuckloch surface mine was shifting some 9 million tons of overburden per annum, in order to extract some 400,000 tons of domestic quality coal each year. It was employing 80 – mainly local – personnel on two 12-hour shifts.

Derek Moir, site manager, explained the challenge faced at Glenmuckloch as the mine clips the Southern Upland Fault, which in some parts of the site has near vertical 3ft thick seams of coal, making it necessary to restrict the overburden removal to 6m high benches for the Cat 325 coal shovels to work in. The coal shovels load a fleet of adapted coal-bodied Cat articulated dump trucks, which make their way along the haul road to the conveyor system. There, a Cat 966H wheeled loader feeds Europe's longest overland conveyor, which enables the coal to be transported without travelling on the public road. ATH estimates a total 2.8 million tonnes of coal will be conveyed, removing the need for approximately 54,000 lorry journeys per year and so greatly reducing the impact of coal transport on both the environment and road network.

The challenging geology makes the available coal 95% recoverable. Parts of the site have been previously worked by deep mining techniques, as witnessed during our time in the cab, as the PC3000 was digging out old wooden pit props – left from

Dust on the haul road is controlled with a number of tractors towing water spraying bowsers.

room and pillar mining – within the blasted rock. At the time of my visit, the mine had reached a depth of 40m with a further 40m to go before reaching the last coal seam.

While the Komatsu PC3000-6 Super Shovel may have a Japanese manufacturer's name on the side of the machine, it appears to be German engineering in Düsseldorf that has won hearts and minds among operational and maintenance staff at the Glenmuckloch surface mine.

Sadly once again, another surface coal mining company was put into administration during December 2012 due to a slump in coal prices and rising costs. However, five months later, in May 2013, all 230 local coal mining jobs had been saved after Hargreaves Surface Mining Limited had acquired 'certain assets' from the liquidator of Aardvark (TMC) – the main trading arm of ATH Resources Plc – for £10.4 million. This allowed coaling and restoration activity to resume at this site, with a Caterpillar D11R, which is covered in detail within Chapter 7.

A modern fleet of twelve Cat 777F haul trucks and one 777D model.

The driver needs to have nerves of steel to track the PC3000 off a 6m high bench!

The PC3000 is sitting on top of old underground mine workings and is loading newly blasted material and old pit props into a Cat 777F.

CHAPTER 4

ATH Resources • PC3000-6 & RH120-E • Netherton • September 2011

ATH Resources Plc was granted planning permission from East Ayrshire Council in June 2010 to set up a new surface coal mine at Netherton and subsequently made an £8 million investment in infrastructure, fixed and mobile plant.

Netherton was ATH Resources' largest and newest operational surface mine, operating a mix of classic RH120-C and the very latest Komatsu PC3000s and Terex – O&K RH120Es hydraulic shovels, Caterpillar and Terex haul trucks and Cat dozers. These monster miners are working together to extract approximately 4 million tonnes of high quality and low sulphur 'black gold' from the Ayrshire coal fields.

This site is located close to the company's original Skares Road Mine, near to Skares village, Cumnock, and employed up to 150 'sunshine miners'. The Netherton site extraction area covered 127ha and was projected to run for eight to ten years, including a residual restoration period of up to two years.

At the time of my visit, coal extraction was well under way with CAT 325C & 320D coal scrapers working the upper and lower seams. The current extraction area is a compact 1km square and an efficient operation, with short haul running to the tip sites and backfill areas. Derek McMurdo, site manager, explained that they are enjoying favourable geological conditions with numerous coal seams close together at a depth of approximately 75m, with the final seam at around 130m and the stratum on a shallow descent, making it easier to extract. The exposed seams were ranging from about 1m thick to as thin as just 35cm. Overburden (which is material sitting on top of the coal) ratios start at 17:1 and increase to an average of 20:1.

At the bottom of the mine is a classic and rare 245-tonne O&K RH120-C backhoe, which originally started

out at the Skares Road mine, and it is teamed with three 91-tonne capacity Terex TR100 mining trucks to haul the remaining overburden to a backfill part of the site. Because the load and haul team are working close to the lower coal seams – and backfilling as they go – this allows very short running, which saves a significant amount of time, fuel and other costs.

The enormous weight of these machines has to be taken into consideration when working close to some of the thinner coal seams. ATH has deployed the RH120-C backhoe to help ensure these fragile seams do not get compacted and damaged as the backhoe is able to dig away from the seam, clearly a technique not achievable with a face shovel. The RH120C backhoe was positioned perfectly on the bench, about 4 or 5m high, to align with the haul trucks' skips to maximise loading efficiency. With its 13 cu.m bucket it is capable of shifting 550–600 cu.m of material per hour.

Once all the overburden is removed, Caterpillar 325C and 320D model coal scrapers are used to load the four Cat 735 adapted coal body articulated dump trucks (ADTs) operated by ATH. During peak coaling extraction, ATH will hire in some additional muscle in the form of a fleet of adapted coal-bodied Volvo A35D ADTs from local plant hire and road haulage contractor T French & Son.

Higher up on the mid-level bench, one of ATH's newest mining shovels, a Komatsu PC3000-6, (call sign Demag four), was teamed with a mix of Terex TR100s and 91-tonne payload CAT777D/F models. The PC3000-6 is capable of loading the trucks in four quick passes. In recent years the PC3000-6 model has enjoyed good sales success in the UK surface mines, with ATH being the proud owner of five of these machines over the last five years. As reported in the

The current extraction area – box cut – is a compact 1km square with short haul running to the tip sites and backfill areas.

Terex O&K RH120-E loading a fleet of Terex TR100 haul trucks, with a TR100 tipping its load a short distant away in the background.

Photographed during a rest break. The ex GM Mining RH120-C shovel was deployed to remove topsoil, which will be stored for restoration of the site.

Cat excavators used to scrape and load the valuable coal.

previous Glenmuckloch visit, both old and new Demag and Komatsu equipment is highly rated at all levels with ATH staff.

Two Terex O&K RH120-Es, one O&K RH120-C face shovel and an ex 1998 prototype Demag H255S are working the upper levels of the site. The former GM Mining RH120-C shovel was deployed to remove topsoil, which will be stored for restoration of the site. One of the RH120-Es was digging out blasted rock and the other machine was working alongside the H255S – the first machine to enter service in the UK, with more details on this later – was tackling unblasted weathered whinstone.

Where blasting is necessary, this activity is contracted out to EPC–UK with the explosive teams working and blasting during the dayshift only, while the two new Atlas Copco ROC L8s drill 7m deep blast holes on a two-shift system to keep the six monster miners fully utilised 24 hours a day, Monday to Friday.

The 290-tonne RH120-Es with their twin engines, 15 cu.m buckets and tri-power design are regarded as some of the most productive machines on site, so much so that ATH ordered a number of Terex TR100s with factory-built spill boards to match the four pass loading capability of the RH120-Es. ATH has also contracted Miller UK Ltd to refurbish some of the older Terex haul truck skips. During refurbishment spill boards will be added; this modification not only allows ATH to maximum a heaped load, but also to reduce spillage around the site.

Whilst ATH claims the RH120-Es are marginally more productive than the other modern prime movers, at 700 cu.m per hour, the Demags are regarded as the most reliable, with high availability rates of 99%. Even the 13-year-old former LAW mining H255S model can sustain long periods of 98% availability and despite its age and smaller 13 cu.m bucket, it is performing within 85% of the newer PC3000-6 and RH120-Es' output and, depending on material density, the H255S will load the 100-tonne haul trucks in four or five passes.

Derek McMurdo, and Ian White, plant manager, both agree that the high availability starts with the simplest of preventative maintenance tasks: pressure washing the machines, and in particular, keeping the cooling pack radiators as clean as possible to keep the hydraulic oil temperature down. Otherwise, overheated hydraulic oil will have a detrimental effect on the hoses and seals. ATH has its own skilled maintenance team on site, with some members of the team factory-trained Demag (Komatsu) Super Shovel experts. The two RH120-Es and 28 TR100 haul trucks are supported by the nine on-site Terex staff, with office and workshop facilities. However, the RH120-Es were reaching the upper limits of their 25,000 hours repair and maintenance (R&M) contract with the dealer.

At the time of my visit, Caterpillar's UK dealer Finning also had staff that were never far away; with seven D9T dozers, seven 996G wheeled loaders, four 35-tonne payload 735 ADTs fitted with adapted coal bodies, a 16H grader, four 777D haul trucks and a fleet of 360 excavators to maintain.

Ian White was pleased to report that ATH enjoys excellent product support from all its external service providers, such as KMG Warrington (Komatsu/Demag), and with both the Terex factory and Finning's Scottish HQ only about an hour's drive away, Ian has all the support in place to ensure the machines are well maintained. In addition to this, along with Glenmuckloch mine, Netherton enjoyed access to imprest parts held at other local surface mining companies operating Terex and Komatsu/Demag mining excavators.

As previously mentioned, ATH Resources gained a RoSPA safety gold award during 2010 for having a safety first approach. With this in mind, ATH had fitted Caterpillar LCD screens and reversing cameras units to its six new Cat D9T dozers. This is a particularly useful visual aid, considering that on some duties these dozers can spend as much time travelling backwards as they do going forward. In addition to this, the offside air conditioning pack had been moved to the cab roof for improved visibility from the cab side window. The veteran Demag H255S has also been upgraded by fitting a large 5.6in colour monitor, supplied by Vision Alert, to cover off the rear and offside blind spots of the machine. Other site safety initiatives include the D9T's ripping and dozing protective segregation mounds for the safety of the drilling rig teams when they are working close to the movement of large haul trucks.

At the time, David Lancashire, group plant director, commented: "The safety of the workforce is paramount to ATH and safety improvements to the plant and working practices are regularly reviewed at the safety

Cat D9T is pictured making a new protection berm for the drill & blasting team to work behind.

A Cat 996G wheel loader feeds the barrel wash hopper at a rate of 175 tonnes per hour, resulting in approximately 50 tonnes per hour yield of good useable coal.

committee meetings attended by the site managers, senior managers and directors. All 57 of the Company's 100-ton dump trucks have been fitted with an additional offside camera and monitor systems to extend visibility to that side of the truck for the operator. Two-way radios are also fitted to every item of mobile plant in the mine ensuring that everyone can be alerted to a safety critical situation."

With the wet Scottish climate to contend with, ATH has a constant battle on its hands to keep the haul road in good order, which mainly falls to the Cat 16H motor grader to maintain the site, supported by the D9Ts. The excavated whinstone material is sometimes used to improve traction on the haul road, but it is a fine trade-off between grip and tyre wear. However, when the going gets really tough on haul road maintenance, only a track-type dozer has the traction to deal with adverse conditions and that is where the Cat D9T dozer is still the weapon of choice for ATH and other mine operators for this task.

In order to ensure every last tonne of available coal is extracted, ATH has set up a barrel wash plant to separate the last of the coal from the other material that is shovelled up during final extraction of the seam. A large amount of this material is stockpiled and a Cat 996G wheel loader feeds the barrel wash hopper at a rate of 175 tonnes per hour, resulting in approximately 50 tonnes per hour yield of good useable coal. A Cat

325C face shovel is used to load the ADTs to transport the washed coal to the stockpiling yard.

The coal stockpiling yard and coal transfer area is situated near the main entrance, where a pair of Cat 996Gs fitted with large coal buckets stockpile and load the coal lorries, along with feeding the crushing and screening plant. Regular sampling of the coal is undertaken to ensure the correct blend is produced, which is mainly used to serve ATH's major power generation customers. ATH also produce coal for domestic home use and is currently exporting crushed and processed coal. The coal is then distributed from the coal transfer area by road, using articulated HGVs that typically carry 28 tonnes per load to the nearby railhead at Crowbandsgate. The mode of transport and transport network used in the final delivery stage would be dependent on the customer, but would in most instances involve final delivery by rail to power station customers via the dedicated rail access, with each train capable of carrying more than 1,400 tonnes of coal.

As explained in the previous chapter, Glenmuckloch visit, during December 2012 due to a slump in coal prices and rising costs ATH was put into administration. However, five months later in May 2013 all local coal mining jobs had been saved after Hargreaves Surface Mining Limited had acquired 'certain assets' from the liquidator of Aardvark (TMC).

The coal stockpiling yard and coal transfer area is situated near the main entrance, where a pair of Cat 996Gs fitted with large coal buckets stockpile and load the coal lorries along with feeding the crushing and screening plant.

A PC3000-6 preparing to load another TR100 with overburden. Note how close the bucket can get to the track frame when the tracks are not directly aligned with the face.

ROC L8 drill machines bore multiple 7m long holes to be filled with explosives for the next blast.

The veteran RH120-C backhoe is still a valuable and powerful tool in this application, with a full bucket of rock in this picture.

The Scottish-built TR100 hauling material to the tip site.

The powerful Terex O&K RH120-E with its tri-power linkage is one of the most productive machines on site.

CHAPTER 5

Demag H225S prototype • ATH Resources • Netherton • Sept 2011

As part of my visit to the Netherton site, I wanted to cover this important machine in some detail. The H255S model at Netherton was the official prototype machine for Demag's 250-tonne class of hydraulic mining shovel back in 1998. However, after February 1999, all new Demag excavators in this class would be rebadged as Komatsu PC3000-1 as the Komatsu partnership with Demag came to an end when the Japanese firm bought the remaining shares and renamed the business Komatsu Mining Germany GmbH (KMG).

The H255S prototype machine was built in 1998 at the Demag factory in Düsseldorf, Germany, and after being one of the main showpieces at the quarry and mining show held at Hillhead, the machine was shipped to LAW mining at its Garleffan opencast coal site in New Cumnock, which had planning consent approved in September 1999 for the extraction of 1.2 million tonnes of coal. The H255S was commissioned on site by a KMG Warrington team of service engineers during October 1999. ATH resources acquired the machine and the Garleffan site with the acquisition of LAW Mining Ltd on 28 November 2003. When the Garleffan site was closed in 2008, the H255S was stripped down

H255S being assembled by KMG Warrington engineers. Image supplied by KMG Warrington.

H255S's first bucket load captured at LAW mining during October 1999. Image supplied by KMG Warrington.

and redeployed to ATH's Skares Road surface mine operation.

Coincidently the same year this Demag H255S was built, ATH commenced operations with its first – venture capital backed – opencast coal mine with ambitious plans to develop the Skares Road site. This proved to be a profitable operation and, with the acquisition of LAW Mining, the group's operations doubled overnight.

At the time of my visit the H255S had 42,000 hours on the clock and was looking clean and in good condition, which is testament to the work put in by the in-house maintenance team. Based on its age, spec-sheet, preventative maintenance regime, and current fuel consumption, here are some extrapolated estimated stats on the H255S. It would have used 92 main hydraulic filters along with 61,000 litres of hydraulic oil, and burned 7.56 million litres of gas oil while operating on a two-shift system. The fuel tank would require somewhere in the region of 8,000 pit stops via ground level access and a hydraulically powered swing-down service arm fitted with Wiggins connections. Productivity wise, based on previous muck-shifting performance of 600 million cu.m per hour, it would have moved more than 24 cu.m of material!

It would appear the Demag design team had got its sums right, as the H255S has evolved over the years into the highly successful PC3000 dash six and on the face of it, looks similar to the original prototype machine. Some of the main differences are the operator's cab, boom, exhaust silencers, hydraulic pumps and connection pins from the centre car body to the track frames.

The most noticeable visual difference between the two machines is the absence of the large 1.5-tonne cab guard, as the new dash six model has the latest Komatsu FOPS Level 2 cab. However, the H255S was well equipped, with the operator benefiting from an air conditioning unit during the abnormally hot weather. Other original equipment benefits are low effort joystick controls, and the foot pedals control the front shovel clam, travel and swing brake operations. As with the new dash six model on site, full instrumentation and electronic text monitoring (ETM) is provided to flag up any performance issues.

The H255's centre car body has a large diameter pin design to attach it to the track frames, whereas the PC3000-6 track frames are bolted for simpler construction with extra rigidity and the newer machine has electronic control and sensing on the three main hydraulic pumps. The older hydraulic pump design may explain why ATH's fuel management data shows the H255S consuming about 10 litres per hour more fuel (at 180 litres per hour) than a dash six model, as both machines are fitted with the same single Komatsu SSA12V159, direct injection, quad turbocharged and after-cooled, 12-cylinder diesel engine, producing an impressive 940kW 1,260hp at 1800rpm.

Given the H255S's high hours, as you would expect, it has seen a number of engine, hydraulic pumps and undercarriage replacements over the years. However, the H255S is still swinging its original bucket, albeit it has seen some forty sets of teeth replaced, and ATH reports it is getting about 1,000 hours out of these components.

Derek McMurdo comments: "Despite its age, Demag 1 (H255S) is still a valuable member of the prime mover fleet and given it has a 13 cu.m bucket, it is 85% productive in comparison to Demag number 4 (PC3000-6) or the RH120-E; not a bad performance considering the newer machines are fitted with a larger 15 cu.m bucket."

The management team at Netherton is keen to hit its productivity targets and, good geological conditions aside, machine availability is king; with the H255S still delivering on this front. At the time of our visit the H255S was delivering 98% availability. However, as with any machine this age, being proactive at replacing deteriorating components is vital to keep this level of availability.

Roger Alexander, general foreman, also rates the Demags, both old and new, and has also worked with the older H185S model. Roger comments: "In my experience the reliability of the Demags is remarkable and they also hold high availability records."

ATH decided to upgrade the H255S's visibility package by fitting a large TFT monitor, supplied by Vision Alert to cover off the rear and offside blind spots of the machine. Other safety modifications are the escape ladder fitted down the nearside front of the machine for a quick exit from the cab should the machine be unfortunate to suffer a fire in the rear compartment, and the catwalk around the cab has been extended to allow the driver to keep the large front window clean.

Komatsu claims the use of castings, designed with the help of finite element analyses, at all major pivot points within the steel structures provide optimised stress flow. And after 13 years of hard digging, John Taylor, maintenance team chargehand, was pleased to report that – despite the high hours – there are no signs of cracking on any of the major structural components.

David Lancashire, group plant director, says of the performance of the H255S: "The Demag is a robust machine which benefits from high quality components throughout which rarely let us down unexpectedly."

We had the rare opportunity to spend a short time in the cab of this monster miner to observe man and machine in action. Given the H255S is the oldest prime mover on site, I was anticipating a less than favourable report. However, long-standing driver Kevin McAtee comments: "I've been driving the H255S for the last four years – out of my 11 years' service – and having driven all the other machines on site, this is the only one I want to drive.

"The 255 is such a stable and comfortable machine to operate, only the PC3000-6 comes close."

Kevin is the first to admit that it's not the most powerful front line machine on site, but feels on balance the H255S holds its own with the newer machines.

Roger reports that, despite the H255S's age, it regularly tackles some of the hardest digging on site, which is located on the top benches, by removing the layers of unblasted weathered whinstone. At the time of our visit, it was consistently achieving four pass loading of the 91-tonne capacity haul trucks in one minute forty seconds.

Driver's view from the cab of the H225S loading a TR100 haul truck. Note the TR100 spill boards to help produce good heaped load and reduce spillage around the site.

CHAPTER 6

Land Engineering Services Limited • Bell 50D • Fife • September 2012

The former Comrie Colliery site is the largest area of post-industrial dereliction in West Fife. A combination of contaminated land, large areas of hard standing associated with the former colliery pit head area, and a 40m high burning colliery waste bing making in excess of 100ha of countryside unfit for any conventional countryside use. The project is to extract 680,000 tonnes of low sulphur coal through working of 52ha of the coal resources, while using revenues from the extraction to accomplish four key objectives. Land Engineering Services Limited (LES) specialises in these types of projects and had three to four years to meet these objectives, which are: to treat the spontaneous combustion and reprofile of the bing (ex-colliery spoil tip); treatment of contamination in Bickram Wood and the former Rexco smokeless fuel plant; and the creation of a new wetland habitat for great newts on the northern end of the site.

LES had invested £4.5 million in fixed and mobile plant to carry out the work, with new large excavators, bulldozers and articulated dump trucks purchased to extract the remaining coal reserves and deal with the reclamation phase of the scheme. King of the hill at Comrie colliery was Caterpillar's top of the range (construction) excavator, a large 523hp, 86-tonne, 390D LME in mass excavator (ME) configuration. The big Cat is teamed with three Bell B50D Mk 7 articulated dump trucks (ADT), at the time; the world's largest ADTs currently available. The 45.4-tonne capacity B50Ds are manufactured in South Africa, with final assembly carried out in Germany.

The new trucks are proving to be very popular with both management and drivers alike, and the latter like the very smooth and comfortable ride. Bell claims this is attributed to the front suspension consisting of independent suspension cylinders that allow oil flow and pressure to be changed constantly to minimise the effects of machine movement. In addition, sensors in the frame continually measure and accommodate for bumps in the surface, while lateral sensors also measure any roll and constantly adjust to accommodate for this. The B50D is also fitted with a fully adjustable air suspension seat, which Bell claims is optimally positioned behind the front axle to help smooth out the ride in tough conditions.

Bell has developed an advanced fleet management system called Fleetmatic that enables operators to monitor their fleet accurately from anywhere in the world – the company claims an industry first, with pole-to-pole satellite coverage – and sends machine data straight to management's laptop or other device. It provides basic data, such as machine hours and average fuel burn, to more detailed information such as individual tipping reports.

From a security point of view, the Fleetmatic system also allows owners to monitor the exact location of each machine, setting virtual geo-fences around their ADTs, and monitor individual driver behaviour using unique driver ID codes that are also used to start the keyless trucks.

Iain Devine, LES director, commented: "I'm very pleased with the performance of the Bell B50Ds. Fleetmatic data shows the average fuel consumption running at 20 litres per hour; that's just 2 litres per hour more than the 40 tonne class ADT on site. Considering the B50D ADTs are carrying 10 to 15 tonnes more per cycle, that's very impressive. The machines have also been reliable with no issues.

"Furthermore, since I'm not on site every day, Bell's fleet management service is set-up to email a report

The Hyundai 800 LC-7A excavator working on the phase four section of the site, loading overburden into a fleet of Volvo A40E ADTs. And coal extraction down in the cut is handled by a Hyundai R290 LC-9 coal shovel and Doosan DX300LC backhoe.

on each machine and I can monitor average fuel burn, the number of cycles per day and tonnage moved, so I can see how productive the operation has been on any given day."

Iain continued: "I've chosen ADTs over ridged trucks for overburden removal due to the nature of this site, which has a number of small tight cuts with steep ramps requiring six-wheel drive trucks to handle the conditions, and with that in mind, the Bell B50D's extra

capacity is the best option to give us the lowest cost per tonne."

With only 500 to 600 hours clocked up thus far, Iain was so pleased with the performance of the Bell B50D he placed an order for a fourth truck, which will come with a new exterior load indicator system fitted as standard to help the excavator operator obtain the maximum heaped load. Iain is in discussions with the Bell dealer to provide a full maintenance package for all four

trucks, which will be delivered through the company's UK service agent.

Coal extraction and reclamation of the site is planned to be carried out in 11 phases. At the phase three section of the site, the 390D LME is swinging a Caterpillar 5.7 cu.m bucket and loading overburden into the B50D's large 28 cu.m skip in five passes. The Cat 390D and B50Ds have exacted material to a depth of some 25m and due to the downward sloping strata, the Cat operator has to position the machine at the same angle to help ensure the counterweight does not strike the

ground as it slews to load the trucks. Due to the thin coal seam at this depth, the 390D skilled operator is carefully extracting the unblasted rock to expose it, which will eventually allow the nearby Hyundai R290 LC-9 backhoe excavator to scrape the coal on these sloping benches and load a fleet of Doosan DA40 ADTs.

Once the Cat 390D is finished loading, the B50Ds face a steep climb on the soft and slippery ramps – due to torrential rain – with which they deal with ease, as they are fitted with massive 875/65 R29 wide tyres. The driver also pre-selects transmission differential locks

The first Hyundai R290 LC-9 coal shovel in the UK loading Doosan ADTs.

to provide an even distribution of torque from the big Mercedes–Benz engine, which produces 2,200Nm at just 1200rpm to each of the six wheels. Once out of the cut, the 50-tonne ADTs make the short haul to the tip site, where they hoist their skips in less than 12 seconds. The discharged overburden is then handled by a Komatsu 85PX dozer belonging to Ward Plant Hire.

The Bell B50D is fitted with a 15.93-litre V8 turbocharged and intercooled engine, the same unit found fitted in Mercedes–Benz road-going heavy trucks. Bell has chosen selective catalytic reduction (SCR) technology to meet the Stage IIIB emission levels. This requires the use of diesel exhaust fluid commonly known as AdBlue, which is an after-treatment injected into the exhaust system and reacts with NOx gases in the catalytic convertor to form harmless nitrogen and water; current consumption of AdBlue is running about 4.5% of fuel burnt.

LES sources its AdBlue from Bell's supplier, and the fluid is available in 10- or 18-litre containers, a 210-litre drum and 1,000-litre IBC. LES currently uses 10-litre containers that have a built in nozzle, and the drivers appear to be quite comfortable using these as they routinely pour one container into the AdBlue tank at the start of the shift. This routine is sufficient to keep the tanks fully topped up. The position of the AdBlue filler cap is a bit of the stretch to reach on the Bell B50D offside walkway/side panel, but surprisingly I did not hear any complaints about this and the drivers just take this in their stride and as part of their daily duties. By and large, each driver operates the same truck, so this helps to overcome the prospect of the AdBlue tank not being refilled. LES is also looking at options to have an AdBlue tank or tanker dispensing unit, which will help in the longer term.

I had an opportunity to catch a ride in the B50D via the standard fitment driver trainer seat and spoke to driver Roger Beverage about its performance. Roger's last truck was a 100-tonne capacity Cat 777D operated by another nearby opencast site, and he was full of praise for the smooth and comfortable ride and air suspension seat of this large ADT. He was also quick to point out the powerful exhaust brake systems and transmission retarder. This provides automatic retardation to slow the truck when the operator backs off the accelerator pedal, which is a useful feature to have when descending the steep ramps and prolongs the life of the service brake components.

The only item on Roger's wish list was a wiper blade system for the large nearside window, as he uses the rearview mirrors to guide the truck into position with the Cat 390D excavator. A welcome and recurring theme on recent visits to new trucks is the low levels of interior cab noise – Bell claims the B50D cab is sound insulated to 76dB(A) – however, drivers are finding it difficult to hear the horn of the excavator without leaving the side window ajar. From an operational and safety aspect, clearly excavator manufacturers, and indeed all other plant original equipment manufacturers, should look to address this issue by fitting louder horns!

The Bell B50D comes with a number of standard safety features, such as the camera system, which provides a good wide-angle view to the rear, and the incorporation of the suspension pitch and roll sensors system that do not allow the skip to tip if the truck is in an unsafe position. This can be preset to a maximum of 18% of roll, and while operating the truck the driver can see the percentage of roll displayed on the main LCD dash panel.

Another 'Top Trumps' machine on site is the 83-tonne Hyundai 800 LC-7A, backhoe excavator – the first LC-7A excavator to be delivered in to the UK market – working on the phase four part of the site loading overburden into a fleet of Volvo A40E ADTs operated by Ward Plant Hire. The Hyundai 800 LC-7A is powered by a 510hp DOHC Cummins QSX 15 turbocharged Stage IIIA engine and is fitted with a standard 4.53 cu.m bucket and with the operator achieving a 110% fill factor, it had the Volvo's 24 cu.m skip full in four big passes and on its way towards the tip site in 40 seconds! Working alongside the big Hyundai is one of Ward Plant Hire's Cat 374 excavators and, again, teamed with the plant hire company's Volvo A40E ADTs.

These two large earthmovers made short work of excavating the adjacent cut, allowing a very rare – we believe the only one in the UK – Hyundai R290 LC-9 with a coal shovel front end to tackle the various layers of coal. At the time of our visit it was dealing with a healthy 1m thick seam of 'black gold'. At the heart of the Hyundai R290 LC-9 beats a 6.7-litre Cummins QSB6.7 six-cylinder, turbocharged, intercooled Stage IIIA engine, producing 227hp.

The face shovel front end equipment was manufactured by Allied Equipment Ltd, and fitted to the Hyundai R290 LC-9 by Hyundai dealer EP Industries Ltd based in Derbyshire. The machine was delivered to site during October 2011 to be the prime coal loader and has been proven to be very reliable during this time. The main advantages of using a shovel configuration over a conventional backhoe is to allow the operator to follow the contours of the coal seam much easier, with better vision for lifting coal at floor level, and with a flat bottom bucket that enables him or her to keep a cleaner product with less wastage. And in addition to this, EP Industries Ltd claims the shovel equipment is a much heavier, more robustly manufactured piece of equipment and designed to stand the rigours of loading, especially when encountering hard seams of coal. The shovel gives better breakout force, bigger bucket capacity, fast cycle times and more control of the product when loading over its backhoe counterpart.

The front end equipment consists of a 4.6m arm, a 3.2m boom and carries a 2.7 cu.m coaling bucket giving a working reach of 9m to load directly from the coal seam into the ADTs. At the time of our visit in 2012, it was loading LES's new fleet of Scania-powered Doosan DA40 ADTs. The new Doosan trucks have SCR technology to meet Stage IIIB emissions and also require the use of AdBlue. The Doosan ADT skips are fitted with a cable-operated tailgate to ensure none of the valuable coal is lost on the way to the stockpile.

Every surface mine is trying to reduce the amount of double handling of overburden by using a cut and fill as you go approach, and as such the land behind the phase four cut had been backfilled and landscaped. However, it was necessary for the ADTs operating in front of the phase four area to haul material to a nearby tip site. There, LES's large 40-tonne, 360hp Komatsu D155 AX-6 bulldozer, fitted with a 9.4 cu.m blade, was making extremely short work of the continuous stream of A40Es, dumping their 32-tonne payload of rock, and it will be reused later in the project during reinstatement.

Over at the phase six dig area we found LES's very clean pre-owned Komatsu PC600LC-8 backhoe,

A new Doosan DA40 ADT heading to the coal pad.

Operator carrying out first use checks around the Komatsu PC600 LC-8. Note the fold down safety access steps to the cab.

A 40-tonne, Komatsu D155 AX-6 bulldozer, fitted with a 9.4 cu.m blade working at the tip site.

Working at phase six dig area is a Komatsu PC600 LC-8 backhoe loading Bell B40D ADTs from Stokey Plant Hire.

manufactured in 2009. This machine is teamed with Bell B40D ADTs from Stokey Plant Hire. The 60-tonne class 430hp Komatsu is fitted with a 4 cu.m bucket and was quickly loading unblasted material – a mixture of mudstone, limestone and sandstone – to reach the first seam of coal, which is just a few metres below the surface. Bob Stevens, LES operations manager, explained that in some parts of the site they start to extract coal at a depth of just 1m below the surface. However, due to the geology of the site, some parts have bowl-like contours to a depth of 34m. That said, the site enjoys an overall overburden ratio of just 13:1. At the time of our visit there was no drilling or blasting in operation. However, when needed, LES subcontracts this work to Ayrshire-based RJ Blasting (Scotland) Ltd.

Operating at the coal stockpiling yard are two 23-tonne Hyundai HL 770 nine-wheeled loaders. The loaders are powered by 8.9-litre Cummins QSL 280hp Stage IIIA engine and fitted with a 4.2 cu.m bucket. The wheeled loaders are capable of loading 28 tonnes of coal into the articulated HGV trailer in five or six passes. Bob Stevens is pleased with the performance of all the

Hyundai machines on site, praising their good reliability, productivity and value for money.

While LES has made a significant investment in new machines, it has also bought a number of other pre-owned machines, such as an 80-tonne reconditioned O&K RH30E backhoe excavator. This is currently sitting on the 'subs bench' at the phase four location and will be fired up and deployed should any of the four large frontline machines have any planned or unplanned down time.

Putting infrastructure in place on any new site is a costly business. LES was fortunate to inherit a large concrete area from the old colliery workings, which is ideally suited for coal stockpiling. However, LES has also bought an immaculately reconditioned Caterpillar 16H motor grader, fitted with modern safety equipment, such as a rear-view camera system, to keep the main haul roads in good condition. The Cat 16H grader was also deployed, along with a tractor-towed Bomag roller, to construct and improve access to the site using some of the red shale material from the adjacent bing.

The new access road was built to connect the coal

Operating at the coal stockpiling yard are two 23-tonne Hyundai HL770-9 wheeled loaders.

A reconditioned RH30E sits on standby at phase four of the site.

Here we see a coal seam just a few metres below the surface.

stockpiling area to the A907 main road. This allows the 44-tonne HGVs laden with coal good access to the nearby Kincardine Bridge over the river Forth and onward through Glasgow to the Hunterston Terminal on the Clyde, where the coal is mixed and blended for various coal usage and customers. The current output of coal from Comrie Colliery is 18,000–20,000 tons per month.

While Iain Devine is satisfied with the performance of most of his new muck-shifting equipment, having had a few months' experience in this application his 'dream team' for the overburden removal would be the three 80- to 90-tonne excavators, teamed with nine Bell B50D trucks.

A Bell 50D tipping its load in just 12 seconds.

CHAPTER 7

Hargreaves Surface Mining • Caterpillar D11R Bulldozer • Glenmuckloch • March 2014

During March 2014 I made a return visit to Glenmuckloch surface mine during the restoration phase of the site, and to visit one of the biggest production bulldozers in the UK – the mighty Caterpillar D11R.

Coal production in the UK is an important source of fuel and forms part of a balanced energy policy for the foreseeable future. Coal currently supplies around 35 to 40% of the UK's electricity generation, which can rise above this at times of peak demand.

Hargreaves Surface Mining Limited has invested more than £40 million of capital expenditure to take over the former Scottish Coal and ATH Resources sites in Scotland in order to start coal production again and to work in partnership with land owners, local authorities and Scottish Mines Restoration Trust to deal with the legacy of large unrestored surface mining sites.

Glenmuckloch was a former coal production site that ceased working at the start of 2013 due to the collapse of ATH Resources. The site has a significant

A ground worker is measuring progress as the D11R lines up to remove the partition walls. Finning UK has modified the standard blade by adding extended side plates and metal plates over the rock guard area to create a 31 cu.m semi-carrydozer blade. Note the slot cut at the top right-hand side of the blade to help the driver judge when the blade is nearly full.

amount of outstanding restoration work, covering both overburden tip sites and a large deep void from previous coaling operations. A small extension to the site had been approved by Dumfries and Galloway planning department and a scheme has been designed for the overall completion of coal extraction and restoration of the site. Agreement has been reached with both the landowner and the bond holder (insurance company) to provide the necessary funds to restart operations. During July 2013 a restoration project was launched by Fergus Ewing, Scottish Government Minister for Energy; the Duke of Buccleuch; Professor Russel Griggs, chairman, Scottish Mines Restoration Trust (SMRT); and Hargreaves CEO, Gordon Banham, at Glenmuckloch.

Buccleuch Estates bought the land at Glenmuckloch back after the collapse of ATH and set up a new subsidiary company trading as Glenmuckloch Restoration Limited (GMR). It has plans for the development of new renewable energy and for future tourism initiatives.

At the time of the announcement, Richard, The Duke of Buccleuch, commented: "This is a real step forward not only for the local community and the estate but also for the Scottish coal industry and the wider energy sector. A solution had to be found that secured employment and could inject some confidence into what is an important industry in Scotland.

"Our discussions with Scottish Government, the local authorities and Hargreaves have been constructive. The outcome is testament to the benefit of a genuine collaborative effort, which bodes well for the future."

GMR has granted Hargreaves Surface Mining Limited the rights for the extraction of coal at another part of the site known as the Lagrae Extension and to undertake its restoration. The new extension started coaling operations during April 2014 and had a target to produce about 15,000 tonnes of coal per month. Work here is linked with the existing area previously worked by ATH, as parts of the site currently requiring restoration are devoid of sufficient soils and soil-making materials. When carrying out excavation work in the new extension Hargreaves will maximise the recovery of suitable soil-making material encountered below the subsoil and this will be placed progressively on to restored overburden profiles.

The project secured around 60 jobs during the restoration phase and the outlined proposals at the extension in the eastern end of the site will help deliver an innovative restoration in the shape of an exposed geotechnical feature where the strata has been lifted vertically (caused by the Southern Upland Fault), which should look very impressive as a visitor attraction going forward.

With some 4 million cu.m of overburden to move or reprofile on the former ATH site, Hargreaves selected the biggest bulldozer from the Caterpillar line-up, the 104-tonne D11R.

This monster machine was pre-owned, having previously clocked up some 22,000 hours in a mine in Estonia. Before the D11R was delivered to Glenmuckloch, the dozer was refurbished at Finning UK HQ at Cannock. The big ticket items on the D11R refurbishment were a full undercarriage overhaul, including a new set of standard 710mm wide track pads. They also overhauled the final drives, repaired various minor oil leaks and completed full 2,000-hour service with oil sampling. Worn parts and repairs were also carried out on the main pin and bearing bores – equalizer bar, blade trunnions, etc.

This D11R was not originally fitted with Caterpillar's telematics system, Product Link, so in order to monitor the machine's vital functions this was installed. This allows Hargreaves to check fuel burn, utilisation vs. idle time, see any machine alerts at a glance and other management information.

As mentioned throughout this book, the UK mining and quarry companies lead the way in fitting additional safety equipment. Hargreaves has fitted a number of important upgrades, such as the fire suppression, which was installed by Kidde, and Spillard's installation of the rear-view camera system and additional mine spec warning lamps – blue flashing strobe lights at the rear, green seatbelt warning lamp and yellow warning beacon. Finning fitted modern safety handrails – similar to that found on the newer D11T model – to either side of the driver's cab, and rear red and white chevrons.

The machine also received a fresh coat of paint, and sharp-eyed readers will have noticed it is sporting the latest Caterpillar 'power edge' logos as well as the up-to-date black-painted engine bonnet, lift cylinders and blade, and it looks just like new.

Standing 15ft 3in tall, 35ft 6in long (ripper to blade) and sporting a 20ft 10in wide blade, the driver eye line is about 12ft off the ground. This is a very big bulldozer and one of only two D11Rs in the UK. The machine originally came from Caterpillar's plant in East Peoria, Illinois, and was configured with a 27.2 cu.m semi-universal abrasion spec blade. However, in order to get the maximum productivity out of the D11R working in this application, and since a D11R in full carry dozer specification is designed to handle a 43.6 cu.m blade, the engineers at Cannock have modified the semi-universal blade into a semi-carry dozer design. This was achieved by fitting 2.3 tonnes of side extensions and extra plating – made from Hardox 450 steel – in front of the rock guard and three layers of Caterpillar – 65 rockwell spec – roll bars welded to the leading edge of the side extensions. The roll bar's round profile is designed to provide maximum wear protection while minimising drag as the edge proceeds through the material. Finning says this modification was not that difficult to design and execute as it has considerable experience modifying standard blades into carry dozer blades for coal stockpiling applications. The team at Cannock has created a 31 cu.m blade capable of quickly transforming the landscape into a more aesthetically pleasing shape.

Slot Dozing

Normal duties for a dozer driver working in the UK mining industry are maintaining haul roads – when the going gets tough – clearing and tidying around the loading area of the haul trucks, and mainly pushing overburden at the tip site. However, that is not what the big D11R was bought for, so with that in mind, the machine was supplied by Finning with operator training and John Blackett from Finning UK's demonstration team has helped the D11R operators learn the art of production slot dozing.

John took the operators through the technique of slot dozing where the dozer starts at the dumping point and works its way back at around 10m during each return cycle. The reason the dozer starts at the dump area first is because this area offers the least material resistance. This then quantifies into reduced fuel burn during the blade load cycle and a lower cost per tonne moved for the customer.

Another part of the slot dozing technique is to sculpt

a trench in front of the dozer, using the trench walls to help retain the maximum amount of material possible during each blade push. By using the walls to retain more material, production can be increased by up to 20% for no additional cost. Also, by adding a negative slope while pushing the material, even more production gains can be made, as for every negative 1% the dozer pushes down, its production is increased by 2% (using gravity as a helping hand). The operator also adjusts the blade angle during different parts of his push cycle.

The first phase is to fill the blade. The operator adjusts the blade to cut, which is a relatively steep angle of attack, and this is used to fill the blade as quickly as possible, balancing pure power with coefficient traction as this is the most expensive part of the total cycle and fuel burn.

The second phase is when the operator has filled the blade completely; he then adjusts the blade angle back to a carry position, reducing fuel burn even more, and slides the material across the bottom of the slot.

The third phase is to release the material at the end of the slot run by tilting the blade forward to help slide the material from the blade with the least amount of effort. When the blade is empty, the operator then selects reverse, which also activates the Cat D11R's Auto Blade Assist (ABA). This is designed to maximise blade positions for the operator's most productive state, so while reversing it adjusts itself automatically back to the first part of the cycle, the blade load position. The operator then repeats the total cycle again, 'load–carry–dump and return' using the ever increasing slot. The training investment from Finning UK will have significant payback for Hargreaves.

Slot dozing has an optimum performance distance. This can be between 60 to 80m in length, but can be extended to 150m before it becomes more cost-effective to deploy a truck and shovel team to move the overburden. Once the first slot has been cut to the required length and depth, the operator will leave a gap around a third the width of the blade – in this case about 2m – as this will create the wall to help retain the material in the adjacent second slot area, and because material overspill at the side plates is reduced to the minimum when working in the slot.

At Glenmuckloch they normally cut four slots before the operator starts to remove the partition walls. This

is most effective by starting at the furthest point from the dump area and pushing the walls into and down the open slot, to help retain maximum blade loads. The operator uses alternative approach angles from the sides, left to right during each complete blade load cycle, which minimizes uneven load on the blade, final drives and undercarriage to help reduce operational costs. By using this technique, once again the outer slot wall helps to retain the material within the blade as there is little or no loss of material over the sides of the blade.

Once the three partition walls have been removed from the four slots the operator will start a new run of slots, working his way across the overburden area until the desired height reduction or contour of overburden material has been achieved. At the time of our visit the D11R was reprofiling on a downward slope, which helps to move some 31 loose cu.m (LCM) of material with an approximate loose density of 1,700kg per cubic metre, or 31 LCM × 1.7 density + 20% increase with the slot dozing technique could see volumes exceeding 60 tonnes of material per blade full.

Hugh Bryce normally drives the newer 48-tonne Cat D9T dozer, however, on the day of our visit, he was at the controls of the D11R and comments: "It's a really good machine and good to drive, has a lot of power and weight which makes it a good dozer and the big 31 cu.m blade can shift some amount of material."

He adds: "The machine has good visibility and shares the same controls as other Cat dozers, which makes it easy to move from one machine to the other."

The Cat D11R is currently operating for two 11-hour shifts. Its 1,606 litre fuel tank is refuelled twice per day during rest breaks, it is regularly cleaned with a power wash and oil levels are checked at the start or end of each shift.

General foreman Roger Alexander was our guide during our visit and commented on the performance of the big CAT dozer: "I'm really pleased with the D11R's performance; up time is running at 99.9% with little or no breakdowns since it has arrived on site, and the only thing that has failed was a warning sensor, which Finning fixed within the hour."

The D11R is fitted with the standard 710mm wide track pads and these were operating satisfactorily during our visit, however, with exceptional rainfall from the start of December through to the end of February,

A ground worker measures the D11R's progress. Note the height of the partition walls as the dozer cuts a deep slot.

the conditions have been tough going at times on the standard track pads.

At the heart of the D11R beats a massive 34.5-litre V8 twin turbocharged Cat 3508B electronically controlled diesel engine, the same engine that is fitted to the site's 100-tonne capacity Cat 777D haul trucks. The engine produces 935hp at 1800rpm and, depending on the type of material being moved, is burning an average

With the 31 cu.m blade full, the D11R is capable of moving material exceeding 60 tonnes and is a highly productive machine in this application.

The Southern Upland Fault will be left as a tourist attraction as it is deemed to be an interesting geotechnical feature. Note the strata rising vertically at the left-hand side of the image.

At Glenmuckloch they normally cut four slots before the operator starts to remove the partition walls and starting the process over again until the required profile is reached. Note the large void and the main offices and workshop in the far distance.

of 115 litres of fuel per hour. Roger reports that some of the boulder clay found in the overburden is very tough on the dozer and impacts on productivity targets; on the other hand, shale material is a much easier push. The D11R is an extremely productive tool and can shift anything from 500 cu.m to 1,100 cu.m per hour depending on the material density and the length of the cut.

Roger comments further: "The D11R dozer can shift a lot of material in a very short space of time. If you come back after a long weekend break, you can clearly see the area it had been working is completely transformed."

The Cat D11R production performance is monitored by a ground worker using Trimble's global positioning system (GPS) and its TSC1 data collector with survey controller can report the levels in real-time with the ability to put the pole down anywhere on site and see an instantaneous grid position, station, offset and cut/fill and pinpoint earthworks progress. The data is sent via the mobile phone network back to base and allows management to check progress on a daily basis.

Hargreaves has acquired from the liquidator all ATH's mining, geological and intellectual property and has retained the services of all its former staff, thereby saving jobs within the local community and acquiring extensive knowledge of the site and the new extension. This has allowed the company to deploy some of the existing prime movers to help with the restoration work, such as an O&K RH120-C face shovel swinging a 15 cu.m bucket and a RH120-E backhoe model – from the former Scottish Coal inventory – in the Bucyrus colours. The backhoe machine is swinging a 16 cu.m bucket. These large prime movers are matched to a fleet of Cat 777D haul trucks, with Cat D9Ts dozers taking care of the material at the tip site. A Komatsu PC3000-6 face shovel is also on site, but out of action having its swing motor replaced.

With a number of mining sites now under the control of Hargreaves Surface Mining and requiring extensive restoration work, the big Cat D11R dozer was expected to be on this site for the remainder of 2014 and then redeployed to start transforming other sites requiring restoration.

First use checks are easily carried out.

There is a reasonably good view to the blade, however fine grading is not required of this machine!

Note the extra safety equipment fitted – D11T style handrails, rear chevrons, fire extinguisher, rear-view camera and warning beacons.

D11R tipping the material at the end of the cut. Note the steep angle of the slope which helps with productivity.

Bucyrus 120E tidying up the loading area as the Cat 777D haul truck pulls away and en route to the tip site.

A refurbished O&K RH120C loading a 91-tonne capacity Cat 777D haul truck.

Bucyrus 120E backhoe – from the ex-Scottish Coal fleet – fitted with a 17 cu.m bucket.

CHAPTER 8

The History and Importance of the Terex Factory • Motherwell • April 2014

The Motherwell factory has been supplying large earthmoving machines through its dealer network since 1950 and the current range of mining, quarry, and articulated dump trucks can be found in most chapters of this book.

The history of the Terex brand began in 1907 in Wickliffe, Ohio, when Armington Electric Hoist was started by George A Armington. The company was renamed Euclid Crane & Hoist when the plant relocated to Euclid, Ohio. It had a number of divisions and one became the Euclid Road Machinery Company (Euclid). This was formally incorporated in 1931 and is credited with having built the world's first purpose-built off-road hauler in 1934 – a 14-ton capacity dump truck called the IZ Trac-Truk. Just 16 years later Euclid expanded its production operation to Motherwell.

The first true off-highway mechanical rigid rear dump truck, the 14-ton IZ Trac-Truk (Euclid, United States). Photo: Terex Trucks

6th September, 1950—The First Euclid

On 6 September 1950 the first truck – an R15 – is made and sold to Wimpey. Photo: Terex Trucks

Due to world demand for Euclid products in the late 1940s, and to grow its share of the sterling market (coupled to the shortage of the US dollar and the devaluation of the British pound), Euclid was looking for a UK manufacturing facility. Archive records show that the site at Newhouse, near Motherwell, was chosen as a prime location due to the availability of excellent skilled labour, good access to steel production – from Colville's (which became Ravenscraig Steel works) in Motherwell – and easy access to ports such as Glasgow and Leith docks that had the capability of handling exports of large earthmoving equipment. This still holds true today as Russia is an important market for Terex, and it ships the trucks in kit form by boat via Leith docks in Edinburgh,

along with Immingham, Hull, Liverpool and Southampton, to reach export markets. From the beginning, the factory also enjoyed close links to both Glasgow and Edinburgh airports for fast parts distribution to its global customers.

The facilities at Newhouse started out under the trading name of Euclid Great Britain and covered 64,000 sq ft. On 6 September 1950, just 51 days after the factory buildings were completed, the first off-highway rolled off the production line, an R15 – 15-ton capacity – rigid truck. This truck was bought by the famous George Wimpey & Co. Ltd and this was the start of an incredible production run for the R15 that lasted nearly 18 years.

By 1968 more than 3,000 of these trucks had been made for home and export markets. A Euclid Great

A restored R17 next to a new TR60 dump truck. Photo: Terex Trucks

Euclid 82-40 just delivered to Shellabear Price Ltd, subcontractors to Tarmac Ltd, on M6 Motorway at Thrimby near Shap, in early 1968. Photo: Keith Haddock

Britain spec sheet from 1955 shows the R15 was fitted with a Leyland engine producing 154hp at 2000rpm, the cab windows were made of shatterproof safety glass and the dump body had 38mm of thick oak planking between the 9.5mm steel top liner plate and the 7.9 mm thick bottom plate. The R15 was superseded by the introduction of the 17-ton capacity R17 dump truck. One such working example was gifted to the factory – serial number 22145 – which will be completely rebuilt as part of Terex's ongoing apprentice programme, starting in 2016. This old truck was last operated by Leiths, a large Scottish quarry operator based in Aberdeen. The last R17 left the factory as late as June 1984 after a good innings covering 16 years in production.

In 1953 General Motors (GM) bought Euclid for 305,137 shares of unissued stock, valued around $20 million, and the Scottish factory then started to trade under a new name of General Motors Scotland Ltd. One year later, the Newhouse factory marked a major milestone as the 1,000th rigid truck rolled off the line. At this time, the factory started to move away from importing large components by fabricating structural components in-house, such as the tipper bodies. In the same year, GM unveiled a range of self-propelled scrapers, the S-7, S-18 and TS-18, at its American Milford proving ground.

In 1957, the TS-18 was updated to become the famous TS24 motor-scraper, fitted with two six-cylinder two-stroke diesel engines rated at 300hp up front and the rear power unit was rated at 218hp. Both engines delivered their power via Allison transmissions to give the TS24 a capacity of 24 cubic yards (struck), or 32 cubic yards heaped. The Scottish factory hit another milestone in 1961 when the 4,000th scraper was produced. One large operator of Terex motor scrapers was George Wimpey (Plant) – clearly a longstanding customer of Terex products – which operated 18 Terex TS24 scrapers, all of which were twin-engine machines featuring a pair of V8-71 General Motors two-stroke engines, on the A13/M25 Tilbury–Southend road construction job in 1980. At the time, this was the largest fleet of Wimpey owned motor scrapers at work anywhere in the UK.

Between 1983 and 1988, the Newhouse factory designed and manufactured 35 TS-8 twin engine scrapes for the British Army. These machines, with a heaped capacity of 8.87 cu.m, were specially made to fit within the hold of a Lockheed C-130 Hercules transport aircraft. The last motor scraper – a TS14G model (20 cu yd heaped) – rolled out of the factory on 30 August 2010, and was shipped to a customer in Canada,

A pair of Euclid R-65s driving from Motherwell to South Wales to one of Sir Lindsay Parkinson & Co.'s opencast coal sites. Photographed at Thrimby on A6 in 1968, before the M6 was constructed. Photo: Keith Haddock

signalling an end to production of the motor scraper products due to a lack of demand.

The year 1968 was a momentous one for General Motors as it divested the Euclid brand (due to anti-competition laws in the US) and the remaining earthmoving division was renamed Terex – taken from the Latin 'terra' (earth) and 'rex' (king). With rigid haul truck size increasing, the factory started to produce a 45-ton capacity R-45. A roll-out ceremony took place on 4 June 1968 when a bigger 65-ton R-65 was 'launched'. With the local MP having a drive around the site, he is reported to have said it was easier to drive than his car!

In the same year, the factory had £3 million worth of orders for the R-65 model, with 27 units sold to an Italian company for the Tarbela Dam project in Pakistan, 12 trucks sold to an iron ore mine in Liberia and 24 sold to UK coal mining companies. The delivery of the trucks to customers today is a much regulated affair under Special Types General Order (STGO) regulation. Back in the

1960s, the big 65-ton capacity trucks were sometimes just driven on the public highway – such as a pair of Euclid R-65s that were spotted and photographed in 1968 (by the well-known author Keith Haddock) driving from Newhouse to South Wales to one of Sir Lindsay Parkinson & Co.'s opencast coal sites via the A6 before the M6 motorway was constructed. This image can also be found on page 68 of Keith's new book, *British Opencast Coal: A Photographic History 1942 – 1985.*

With a booming order book, the factory management commissioned an extension for a new assembly line, which was officially opened in 1970. Around the same time the factory announced the purchase of adjacent land for a new 50 acre £1.8 million parts centre, which freed up production capacity, with the manufacturing site now covering 120 acres. During construction of the parts centre, a factory-produced 275hp 82-40 model bulldozer was used by the main contractor, George Wimpey, for general site clearance duties. Another

Euclid 82-40 bulldozer from the factory was delivered to Shellabear Price Ltd, subcontractors to Tarmac Ltd, on the M6 motorway at Thrimby near Shap, in early 1968. It was photographed by Keith Haddock as he was helping to build the road back then.

At that time (1969), the number of employees reached 1,611 at Newhouse with some additional 150 in the pipeline. In the same year, the 33 Series haul trucks were introduced, including the world's largest truck, the Terex 33-19 Titan (built in the USA). This single Titan prototype was in service until 1990 and is now on display in Sparwood, British Columbia, near the mine it served. At 370 tons, until 2013 it was considered to be the world's largest haul truck by dimensions, but not by carrying capacity (Caterpillar's 797F — from 2009 — could haul 400 tons but was smaller in size to the Titan). Belarusian company Belaz launched the Belaz 75710 in October 2013 to fulfil an order by a Russian mining company and it is currently the world's biggest dump truck. It has the capacity to haul a massive 496-ton payload and stands at 20.6m long, 8.16m high and 9.87m wide.

In 1970 the Newhouse factory won a Queen's award to industry for the second time, having first received this prestigious accolade in 1964, the year of the award's inception. During the 1970s, GM's Terex Division created some of the industry's most notable heavy construction equipment, including the world's first twin-powered dozer (TC-12).

In July 1973 GM announced a massive £3.3 million order for 112 of its 82-30 dozers, rated at 225hp, for the state of Goias in Brazil. The major proportion of the order was supplied by the Newhouse factory, with the remainder coming from the Hudson factory in the US. In the same year, GM ambassador and famous golfer Jack Nicklaus visited the factory and entertained the staff by hitting shots from a 'tee' on the top of a Terex TS-24 scraper box.

While the team at Newhouse built the products; it was Blackwood Hodge that sold the machines during the early years. This dealer was formed in Scotland by John Blackwood and Neil Hodge in 1941, trading as John Blackwood Hodge & Co Ltd. It was bought by Barnard Sunley, with its HQ now based in Northampton, and by 1953 it became a publicly listed company, Blackwood Hodge. In 1978, on its 25-year anniversary, Blackwood

Hodge had grown into the biggest earthmoving equipment dealer in the world, with 6,000 staff in 30 countries and had also grown worldwide sales from £2 million in 1953 to £60 million in 1976. Today, TDL Equipment is the official UK dealer of Terex heavy line construction equipment, Atlas material handlers and Genie compact telehandlers. In 2012, TDL centralised all its Terex, Atlas and Genie parts businesses by moving them to Tankersley, near Sheffield, for good access to the M1 motorway network. However, some TDL staff still work out of the Newhouse factory, which has a large centralised stock of parts.

A range of wheel loaders were designed and built at Newhouse. The smallest was a pivot steer 72-11 model with a 1.15 cu.m capacity powered by a 94hp four-stroke six-cylinder engine with a distinctive single lift arm. Its production run was short-lived, with the first one produced in May 1971 and the model discontinued in 1975. At the other end of the scale was the flagship 72-81 model — sporting a 7.65 cu.m bucket — which started out life under Euclid, in January 1968, as the 72-80 and was launched under Terex in 1969 as the 72-81. It was powered by a V12 two-stroke diesel engine rated at 465hp. The machine was upgraded in 1973 as a B model and featured the Terex trademark keystone grille and stayed in production till 1982, when it was replaced by the Terex 90C model. Having lived near the factory since childhood, I was thrilled to be invited by a friend and member of staff to an open day visit in 1989. I watched a display involving a 90C loading a 3310E haul truck — a good and memorable show!

During the visit we spoke to long-serving staff; Frank Connelly started working for Terex in 1970 as the first structural apprentice and had been working as a structural engineer there for over 44 years. However, industry and Terex have moved on since the 1970s and Frank maintained that these changes were for the better. New technology has enabled a higher quality product with the use of plasma cutters, which replaced profile burning. This change minimises dust and corrosion and maximises precision. Gone are the days of welding using a stick and rod, as the structures are welded together using the latest MIG machines and some robotic welding is also carried out. Terex invested £4.7 million during 2012–13 in manufacturing equipment. Of this, £3.3 million was spent on three new CNC metal-cutting milling machines to take control of this process back in house.

Changes in quality control have meant that ultrasonic machines are used to ensure high standards of fabrication are maintained. Frank explained: "The new machines are 'fantastic' compared to what they were like years ago. The ongoing business investment into these new pieces of technology is really welcome and provides confidence for the future." Frank also mentioned that the best investment for the company lies in the apprenticeship programme, which means that each year approximately six more people start their working life building earth-kings of the future.

On 1 January 1981 GM sold the Terex Division to IBH Holding AG, with the Terex division becoming Terex Corp, a subsidiary of IBH. GM retained a small shareholding in IBH and remained the freeholder of the Newhouse plant, which turned out to be very important for its future. In November 1983, due to a downturn in the global economy, the German bank that was financing IBH collapsed and took IBH with it. However, GM was not going to let its former earthmoving division get cut

up under the liquidation of IBH assets. As the freeholder of the Newhouse factory, it proved a key player in saving the Terex brand by repurchasing Terex Ltd (Scottish operations) on 19 February 1984 with the new trading name Terex Equipment Ltd – a subsidiary of GM. For the staff at Newhouse this meant an eight-week break in service between the liquidation of IBH Terex and the rise of Terex Equipment Ltd.

A change of ownership came again in December 1986, when Northwest Engineering, an excavator and crane manufacturer, purchased Terex Corp and assumed the Terex Corp name with an option to purchase Terex Equipment Ltd. This it exercised on 30 June 1987. By 1997, Terex Corp had established two business divisions, Terex Trucks and Terex Cranes. Terex Trucks included all earthmoving, construction and mining equipment companies, including Terex Equipment Ltd, which would be the world headquarters for Terex trucks, based at Newhouse.

One of the major achievements of the Newhouse factory was the design and manufacture of the first ever Terex articulated dump truck (ADT) in 1983. This was

Now transported on the back of a heavy haulage low-loader trailer, pulled by a powerful HGV articulated vehicle, this TR100 was heading for Broken Cross coal mine.

The skilled workforce at the factory turn sheet metal into high quality rigid haul trucks.

A 91-tonne capacity TR100 nearing the end of the production line.

A specially adapted waste body was designed at the factory and the order placed with a TDL equipment customer within two weeks!

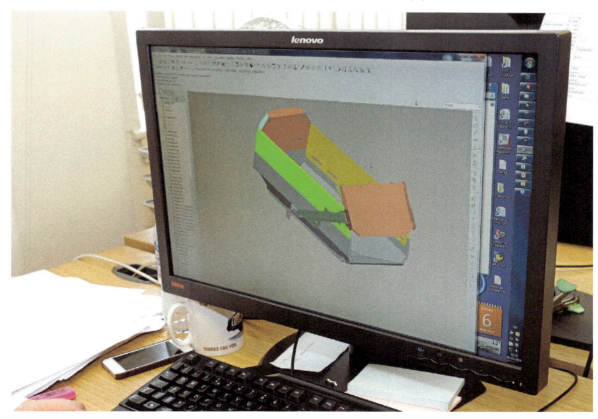

Image overleaf courtesy of Terex Trucks

One of the first ADTs designed and built at Motherwell, a 23-04 model. Photo: Terex Trucks

arguably one of the most successful product lines in recent times, as the production split in 2013 was 62% ADT-built trucks compared with 38% rigid bump trucks. The first model designed was the 23-04, a 23-tonne capacity truck, powered by an air-cooled six cylinder Deutz engine producing 212hp connected to a six-speed ZF gearbox with auto lock-up – on the torque convertor – in the top four gears, driving the front and foremost rear axle only (4x4). In 1985 this designation changed to 23-66. This stands for 23-tonne and now three (6x6) drive axles.

In 1998 the two truck lines changed from their famous green livery to white and the TA (Truck Articulated – ADTs) and TR (Truck Rigid) lines were introduced. In 2002 the Generation 7 ADT was produced and in 2005 Newhouse produced its 1000th ADT, a Gen 7, 30-tonne, TA30 model. Four years later the Generation 8 truck was produced and the model numbers were changed to TA250 (25 tonnes), TA300 (30 tonnes) and TA400 (40 tonnes). The year 2011 heralded the introduction of the Generation 9 ADTs powered by

Scania Tier 4 interim/Stage IIIB engines and in 2016 the branding on the trucks was changed to Terex Trucks. Also in that year, the TA400 Gen 10 ADT was introduced in Munich at the Bauma trade show. There are eight models manufactured in Scotland, with a strong heritage, technical expertise and modern process for the following range of trucks:

TA range of articulated dump trucks
Three models – 25 to 38 tonnes
TR range of rigid dump trucks
Five models – 32 to 91 tonnes

On 9 December 2013 Terex Corporation announced it had agreed to sell its truck business to Volvo Construction Equipment for cash proceeds of approximately $160 million. Included in this transaction was the manufacturing facility in Motherwell.

At the time of the announcement, Ron DeFeo, Terex chairman and CEO, commented on the deal: "The truck business has been an important part of our company for

This is believed to be one of the last motor scrapers, a TS14G model, which was rolled out of the factory on 30 August 2010 and shipped to a customer in Canada.

A 91-tonne capacity Terex 33100 (which became the TR100) being loaded by a Cat 5130 at Drumshangie OOCS, Airdrie. Both machines – and the other 5130 in the background – are owned by Bridgetown Mining. Photo: Nigel Rattray

more than three decades and continues to produce world class products with dedicated and talented employees. However, trucks no longer fit within our changing portfolio of lifting and material handling businesses. I am confident that the truck business will benefit by joining a company sharing similar competencies and offering complementary products and services. We are pleased to have entered into this agreement with Volvo, which represents a strong strategic buyer for the business who values our distribution network and team members."

Commenting on the rationale of the deal, Volvo CE's former president, Pat Olney, said: "This is a strategic acquisition that offers Volvo CE considerable scope for growth. The addition of a well-respected range of rigid haulers extends the earthmoving options for customers involved in light mining applications at a time of renewed confidence in the sector. The addition of TEL's articulated hauler range will enhance our position in this segment, particularly in high-growth markets. We believe that the Motherwell facility and its global team members, as well as the current distribution partners, are valuable to the success of the business in the future."

One of the development areas that have taken up valuable R&D resources, – along with improvements in the rigid trucks and Gen 9 &10 ADT cabs – is the ever changing legislation on engine emissions.

Paul Douglas, director and general manager, joined Terex in 1986 as a graduate engineer working on the design of a special 71-72B wheel loader for the British Army. He then left Terex in 1988 to work for Komatsu in Newcastle, returning in 2005 to head up engineering. Through a career development programme, Paul took up his current post in late 2011. Paul was born only 10 miles from Newhouse, so he has a close connection with his Scottish workforce. It is evident that Paul and his colleagues are positive about the future and at the time of our visit he commented: "We are just about to embark on a journey with our new owner, Volvo and, given the current climate, this is one of the best things that could happen for this business. I'm confident that the two ADT products from both Volvo and Terex will

successfully carry on, as our customers will understand the differentials between the two product lines."

Paul added: "With regard to the rigid line of trucks, this is very attractive for Volvo, as they don't currently have a rigid truck offer, and while our TR range of trucks have the latest engines and transmissions fitted, the product line would benefit from investment and Volvo have an appetite and passion for engineering and innovation going forward."

With a global customer base, Paul tours the world and hears a consistent message from his customers, which is that they want safer, more comfortable trucks so their drivers are more productive, along with the lowest cost per tonne moved and lowest cost of ownership.

For the foreseeable future, both Terex and Volvo value the Terex dealer network and wish to continue to support them. As part of the deal, Volvo can use the Terex brand for five years with an option to extend this arrangement in the future. In parts of the world where there is no Terex dealer present, but Volvo has one, then clearly there are opportunities to sell more Terex rigid trucks under the new Volvo arrangement.

Paul Douglas and his team are very customer focused. As an example, a truck rear axle needed to be shipped urgently to the USA, so they took that component off the production line, which had an impact on productivity at the factory, but this kept the customer happy and supported.

Shop floor staff and management alike recognise that the construction industry has been going through various boom and bust periods throughout the company's history and while the team sometimes faces a shorter working week to reflect demand, the workforce is optimistic about the announcement by Volvo about buying the business. They feel Terex is still a great place to work, overall quality is still high and investment is significant. And while the Motherwell factory has had many owners over the last 64 years, the future looks bright and, come what may, one thing is clear: there is a great deal of optimism from management and staff at the factory.

CHAPTER 9

Banks Mining • Rusha surface coal mine • West Lothian • July 2015

In January 2012, Banks Mining, part of the Banks Group, commenced preparatory operations at Rusha surface mine, which is situated near the village of Breich in West Lothian. This coal mine created around 50 jobs, with the preparatory works including the construction of the site compound and offices and development of two new bridges over large diameter water mains at a cost of £1 million. Initial removal of soil and placing into landscaped screening mounds also took place, followed by the main excavations six months later.

Banks has planning permission to undertake operations at Rusha for a seven-year period and at the time of my visit the mine had been in full swing for three years, having extracted more than 750,000 tonnes of good quality low sulphur coal in that time, and the Banks team expects to extract a further 250,000 tonnes before commencing restoration of the 154ha site. Banks Group has been in business for 40 years, and has a proud record of fully restoring all 110 sites that the company has operated. The Rusha site will be fully restored and landscaped to its original state with a mixture of woodland and agriculture. Banks also has an exceptional track record of designing and fitting noise reduction technologies to its heavy plant that helps to minimise the impact of operations on the local community.

I was shown around by site manager Ian Ritchie and assistant manager Cameron Gibson. Our first stop was at Section B, where the company's pristine 320-tonne Caterpillar 6030 face shovel was working hard on a new cut to remove boulder clay overburden. Ian explained that this material is extremely hard going, even for this modern 'sunshine miner' with its legendary tri-power boom and dipper design, and the ESCO teeth on the 15 cu.m bucket need replacing about every eight weeks so it can cut through this material effectively. At the top of the bench was another Cat machine, a 50-tonne D9T dozer ripping the top surface where the majority of the large boulders were being found, easing the load on the big face shovel.

At the controls of the 1,530hp Cat 6030 was Grant MacLanachan, who has been the lucky operator of this monster miner from day one, and I was invited up to the cab to see him put the excavator through its paces in these tough conditions. With safety in mind, the three-year-old Cat 6030 – with 5,900 hours on the clock – has a hydraulically actuated set of access steps fitted as standard and they can be lowered in an emergency via a nitrogen accumulator, which ensures that the steps remains operational even when the engines are shut off. This design is one of the best I have seen on a 300-tonne class machine, as it has a reasonably shallow angle; it's just like going up a flight of stairs at home!

The 6030 has a very spacious cab, where large safety glass is used for all the windows. For extra safety the front windscreen has armoured glass, and the cab design provides an excellent all-round view from its lofty position at 6.5m high. The cab also has a trainer seat to bring new operators up to speed. The driver has a comfortable air suspension seat and the light electro-hydraulic servo control joysticks are built into the chair and adjustable to suit an operator's preference for feel and response. Adjustments to the controls are made via a large 12in colour touch screen LCD display, which is mainly designed to monitor machine performance. However, it also has an operator's manual, and even a parts book is stored within its Windows-based operating system and menu. Also with safety in mind, Banks Mining has fitted two rear-view cameras and screens, one to cover the rear of the machine and the other to provide a good view down the right-hand offside blind spot.

Grant is a highly experienced operator, having worked at a number of different surface mines, and has operated slightly bigger 310-tonne class machines in his time. However, he feels nothing in this size of machine will out dig the big Cat 6030. Grant explains: "Boulder clay is really hard on the machine and bucket, but the 6030 is the best machine I've driven and copes well in these conditions as it's got so much power from the two engines and the tri-power system makes such a difference to the digging performance compared to any other machine I've operated." Despite the difficult material, Grant operated the 6030 to four pass load the 91-tonne capacity Komatsu 785-7 haul trucks in about one and a half minutes.

This Cat 6030 is believed to be one of the first in the UK and is certainly the first machine in Scotland. It was purchased when Caterpillar had just completed the buyout of Bucyrus. It was delivered with the Caterpillar logos, a Cat yellow colour stripe and the rest of the machine was ordered in white to match some of Banks' other prime movers from the Terex O&K era of ownership of the RH series machines. While the 6030 is sporting Caterpillar livery, its DNA goes back to the legendary O&K RH120 series of machines, and in fact this new machine still has O&K logos on the travel pedals. Readers will have noticed the large louvred grilles on the rear-mounted engine radiators and cooling packs on the offside front; these modifications are there to help suppress the noise from the large fan blades, and this technology is used extensively at other Banks surface mines, both on the excavators and haul trucks where required.

Ian comments on the machine's performance: "The 6030 is three years old now and we can't fault it as it's not had any down time and is never any trouble. It also consistently meets its production targets, so you can't ask for any more than that."

The 6030 is fitted with two engines, each with their own hydraulic pumps, which when compared with competitor machines in this class gives it an advantage should one of the engines develop a fault, as a twin engine machine can keep working, albeit at a reduced rate of knots. With one engine it can also be tracked away from the face for maintenance and would not hinder any ongoing operations.

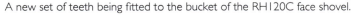

A new set of teeth being fitted to the bucket of the RH120C face shovel.

A refurbished RH120C backhoe is a key prime mover at this site as it is tasked with exposing the old mine working and the hard rock just above the coal seam. It is fitted with a 13 cu.m bucket to load the 100-tonne trucks in five quick passes.

Banks also operates two of the original RH120C machines on this site; both have been refurbished and are in 'as-new' condition. At the time of our visit, the RH120 face shovel was having a new set of teeth fitted by a local welding firm, as the old teeth are cut off by the 'gas-axe' to protect the removal of the locking pins on the teeth adapters.

The RH120C backhoe, with 35,000 hours on the clock, had a reprieve from being a giant parts bin, as it was parked up at the Banks Group main plant depot in Thrislington, County Durham, and was being used for spares before the decision was made to put it back to work with a new set of engines, exhaust systems, hydraulics and undercarriage overhaul. With an additional 5,000 hours now racked up it is an invaluable tool on this site, claims Ian. The 120C backhoe, fitted with a 13 cu.m bucket, was operating at about 34m below the surface and the site enjoys a good overburden ratio of just 16/18:1, which means moving 16–18 cu.m of overburden for every tonne of coal produced.

Ian also explained that the site has some areas where it had previously been deep mined and there are no historical records of the old mine workings. The big 120C backhoe is ideally suited to dig down in front of the heavy plant to find any old workings, although on some sites a smaller excavator could be deployed for this type of work. While Banks Mining has planning permission to carry out blasting, and did blast in Area A, the operations are proceeding quite well without the need for this at the moment, relying on the sheer grunt of the big Cat 6030 and the two O&K machines to break out the rock.

Working alongside the 6030 prime mover was a Cat 824C wheel dozer, made in 1987, affectionately known as 'the bike' and sporting the classic Caterpillar logo. This type of machine is used to quickly get to and from the opposite ends of the site to clean up at the loading area, which helps to keep tyre damage down to a minimum on the haul trucks. Like so much of the kit at Rusha, it is a testament to all the staff that every machine old or relatively new is in near mint condition, and this wheel dozer looks in outstanding condition for its age. And it is clear to me that everywhere you look within this mine, Ian and his team run a very professional operation.

Boulder clay – hard and abrasive digging but the Cat 6030 is built for it.

Grant at the controls of the 6030 keeps one of the cleanest cabs I have ever seen in an opencast site. The Cat 6030 has a 12in machine monitor and two cameras to cover the offside and rear of the 300-tonne face shovel.

Another D series Cat ADT converted to a servicing lube unit and large clean maintenance workshop in the background.

Cat 6030 fitted with a 15 cu.m bucket and is four pass loading in one minute thirty seconds. Once loaded, the 6030 and RH120Cs have blue LED lights to let the truck driver know to move off. Banks does not use the machine horns for this to help reduce noise on this site.

Immaculate Cat 824C 'the bike' wheel dozer tidying up the loading area, while up top is a Cat D9T ripping the top of the boulder clay to break it up to assist the Cat 6030 face shovel.

Banks has its own plant maintenance engineering team on site and operates out of an immaculately kept modern workshop, which helps to maintain all the equipment to a high standard. Servicing of the Cat 6030 is undertaken by Finning UK, and the Komatsu 785 trucks are looked after by Marubeni–Komatsu ltd.

Once the coal is reached, Banks has a mix of two Volvo EC223D backhoe machines and a new Volvo EC300D with coaling front end fitted to scrape and shovel every last piece of valuable 'black gold'; into a fleet of 40-tonne class Volvo A40Fs and some older 40-tonne class Cat 740 articulated dump trucks (ADTs) – both of which are fitted with coal hauling bodies with rear doors. Ian explained that the coal seam, where the Volvo excavators are working, contains an awkward 200mm thin band of hard interburden between each of the 0.5m thick seams, which prevents the coal shovels extracting the coal in one lift.

There are also Volvo and Cat ADTs with standard muck shifting bodies working with a 70-tonne Cat 365L ME (in Mass Excavation spec) which is used for soil stripping. It is considered to be a really good machine for this type of work, as it is fast, powerful, reliable and a good match for the ADTs. During my visit the Cat 365 was making short work of clearing the topsoil by sitting on a 3m high bench to get the maximum digging performance out of the machine to load the waiting trucks.

The other new kit at Rusha are six Komatsu 785-7 91-tonne capacity haul trucks; the 785-7 trucks were built by the Hamilton branch of Marubeni–Komatsu Ltd at local heavy haulage firm Cadzow before it transported them to the site back in September 2012.

Ian is delighted with the performance of the Komatsu haul trucks, praising their good fuel efficiency and reliability. Plus, the drivers love them as they feel they are one of the most comfortable trucks on the market. The front suspension of the 785s are fitted with a Macpherson strut and lower wishbone design, the same layout that you will find in a modern car. The only difference is the huge size of the component to survive the rigours of surface mining operations, particularly in Scotland where the ground conditions can be a challenge due to the inclement weather!

Volvo EC300DL fitted with special coaling shovel front end equipment.

Volvo EC300DL coaling shovels and A40F ADT adapted coal body trucks being loading with high quality coal.

Cat 365C LME on topsoil removal duties at C section of Rusha mine.

Ian is also pleased with the level of management information he receives, via Komatsu's Komtrax system. In an example of this, one of the trucks was recently identified as burning slightly more fuel than the other five trucks, and on further examination of data coming from the truck, it was found to be taking four seconds longer to raise its hoist cylinders to discharge its load. This was traced to an underperforming hydraulic pump, which was quickly changed, thus preventing unplanned down time and possible damage to the rest of the hydraulic system. The skips on the Komatsu 785 trucks have been modified to Banks' own specification – front spill board side plates are modified, nearside and offside side deflectors are fitted and a longer tail chute is fabricated to enhance load ejection.

Banks operates two Cat 777Ds at Rusha and these older trucks are fitted with engine encapsulation side panels, enlarged radiator cowling – with an acoustic splitter – and exhaust suppression kits, which reduces the noise by 17dB(A). Put in context, a 10dB(A)

reduction produces a perceived noise level of 50%; these modifications were designed and fitted by Banks to keep the noise within the site's environmental plan. The newer Komatsu 785-7 trucks are quieter in standard spec; however, Banks has also fitted an extended radiator front cowling to allow louvres to be fitted to reduce the noise. Another noise reduction initiative is large blue LED lights fitted on the 6030 and RH120Cs. These let the truck drivers know when to move off, when fully loaded, as Banks does not use the machine horns for this traditional signalling method at Rusha.

During my visit I hitched a ride in the trainer seat with Andy Penman, who drives one of the 785s, fleet number D299. He finds the 785-7 to be a nice comfortable truck with everything a driver could wish for as they are quiet, fitted with air conditioning, and have great all-round visibility along with a good heated mirror package. The main driving lights are powerful, which makes a significant difference when operating in reduced winter light.

Andy explained that the 785 has a good traction

Komatsu 785-7 captured at Shotts on route to Rusha mine in 2012.

Komatsu 785-7 being built in Cadzow yard. Note the front suspension is a MacPherson strut with a lower wishbone design to provide good levels of ride and driver comfort. Management are very pleased with the operational performance of these trucks.

The RH120-C backhoe in the foreground and the new Cat 6030 in the background loading Komatsu 785-7 trucks. On the right-hand side you can just see the Cat 365LME soil stripping and on the left-hand side – just out of shot – is the coaling operation area.

control system – via an ASR switch on the dash – and it also has a power-boost button that delivers up to 10% extra grunt from the 30.5-litre V12, 1,200hp Komatsu engine when climbing on the haul roads.

The other Komatsu on site is a rental D61PX dozer fitted with a PAT blade. This has been brought in to work with the Cat 365LME and ADTs and it spreads and shapes the topsoil that is being stockpiled. It will also be used during the restoration phase of the operation.

Helping to keep the haul roads in tip-top condition are two Cat 16H motor graders and when the Cat D9Ts are not working at the tip sites, they are used to spread good quality sandstone material to help obtain a balance between traction and tyre life when the haul roads get wet and muddy.

The mine has a number of older ADT machines that have been converted for other uses. Two Volvo A35 models have been converted into 12,000-litre capacity dust suppression units to keep the fine dust under control on the haul roads and truck loading areas. There are also

two D series Cat machines; one has been modified to carry all the different hydraulic, engine and transmission oils by using IBC containers on the back of the truck. While some surface mines favour an agricultural tractor and trailer bowser for refuelling all the equipment, Banks' machine of choice here is an industrial-sized bowser, in the form of a Cat D series ADT, which has a modified ADT body shaped into a 9,000-litre capacity fuel tanker that keeps the prime movers' tanks topped up twice a day during rest breaks.

While Banks has made a significant capital investment in the latest earthmoving equipment at Rusha, it has not forgotten to invest in the people who operate it. All the plant operators have completed a Level 2 NVQ following on from health and safety awareness training and all have taken part in a practical IT training course. The training was carried out by a local training company, Edutrain. Banks recognises that its employees are trusted with millions of pounds worth of plant and also knows the importance of maintaining operator competence in a safety critical industry. Since modern equipment is changing, with technology playing an increasingly significant role, this training helps to ensure operators have a good understanding of information communication technology and Ian has seen an increase in minor near-miss reporting to help improve safety.

Ian pointed out that it is of great importance that the operational team is fully confident, competent and capable of operating to the highest standards. 'Development with care' is Banks' mining ethos, which extends to every employee and encourages them to take great pride in operating in a safe, responsible and efficient manner to ensure they have a safe and rewarding place to work that meets its customers' needs.

After a two-year trial at the Rusha site, Banks has made a six-figure investment in the latest fuel monitoring systems by working with fuel management specialist Liquid Management Solutions (LMS) to help improve costs and the environment with reduced emissions from its plant fleet. LMS provides Banks with real-time data on every aspect of the fuel supply chain across its three active surface mines. Usage by secondary items such as lighting towers, pumps, generators and subcontractors is also captured via alternative options to provide a complete analysis of fuel usage, which can then be used to highlight areas for improvement.

Jim Donnelly, operations director at Banks Mining, is pleased with the outcome of the work that Liquid Management Solutions has done, which is enabling the fuel use on site to be more closely monitored and controlled, and thus help to find better ways of working by reducing emissions and controlling costs.

Mining companies are rightly focused on managing costs; however, Banks is proactive in supporting local community groups and voluntary projects via its community trust fund, where 10 pence is set aside for every tonne of coal extracted. That amounts to £100,000 based on actual and projected tonnage. Banks runs regular poster campaigns to ensure community groups are reminded that funds are still available.

It was clear from my visit that Banks Mining is keen to provide jobs for, and work with, the local community and businesses. This includes welding, training and the latest IT systems, while safety and keeping noise to a minimum are also key priorities. Banks estimates the Rusha surface coal mine will contribute more than £5 million to the local West Lothian economy every year through wages and other supply chain contributions. All the coal extracted will be used by Fergusson Group to fulfil supply contracts in the UK energy sector, of which, coal production still accounts for around 35–40%.

Throughout this book's mining section the aspect of restoration in surface mining has been covered. In some cases it is being put at risk due to the operating company going into administration. With that in mind, I received an invitation to look at just two examples of Banks' fantastic record of restoring all 110 sites it has worked. And while this book is about earthmovers in Scotland, I make no excuse for hopping just over the Scottish border to see some great examples of restoration work: at Brenkley and the former Delhi coal mine, and an opportunity to see a very special piece of restoration work at Banks' biggest surface mine at Shotton. This is notwithstanding the opportunity to also visit the largest hydraulic excavator and dump trucks operating on these shores, which I will cover in the next two chapters.

CHAPTER 10

Banks Mining restoration visit to Delhi site and Brenkley Lane coaling operations • September 2015

Following a visit to Banks Mining's Rusha surface mine in central Scotland, I was invited to see the restoration work carried out at its Delhi site and to visit the coaling operations across the road at Brenkley Lane, where progressive restoration work can also be seen.

Banks Group is proud of the fact it has restored all 110 surface mines worked since the company was formed 40 years ago, in 1976, as its company ethos is 'development with care'. Starting back in 1977, Banks Mining has been working on 14 different sites on Lord Ridley's Blagdon Estate, Northumberland; these include Delhi, the current operations at Brenkley Lane and the massive Shotton surface coal mine – just a short distance away across the A1 – to name but a few.

The restoration of the sites on the Blagdon Estate has included some 40ha of broad leaved woodland and 160ha of agricultural land, which has been divided up by planting new species of hedgerows and hedgerow trees. Banks has landscaped the former mines to include fishing lakes and wetlands, along with ponds, water courses and wild flower meadows. Banks has also provided funding to Northumberland Wildlife Trust to carry out vital works to Prestwick Carr site of special scientific interest (SSSI) close to the former Fox Covert surface mine, which is just a few miles from Brenkley. The aim is to make a significant improvement to biodiversity as well as linking important habitats across this part of Northumberland.

Banks goes to considerable lengths to help ensure the land is restored to a very high standard. In one such example it carried out extensive research into historic plans and aerial photographs, which led to the recreation of a Capability Brown-style landscape that had been destroyed in the 1950s by post-war mining and the drive for food production. This long-running partnership with the Blagdon Estate has given it the opportunity to develop a wider scale landscape. Owen Paterson, then Secretary of State for Environment, Food and Rural Affairs, opened 51ha of the restored land at the former Delhi surface mine, at Blagdon Hall, in 2013.

My tour guide for this visit was Darren Banks, the site engineer. Darren is responsible for planning the excavations, managing the drilling and blasting contractor and for the environmental and geological challenges faced in surface mine operations at Brenkley. Banks Group is a large family business; Darren is the grandson of the chairman and company founder Harry Banks OBE, and works closely with his father Kevin, site manager at Brenkley. The next generation of this family-owned company are clearly a hands on management team.

As mentioned in the chapter on Rusha operation, Banks Mining kit is always found in immaculate condition, with no change to that high standard here at Brenkley. This is down to the hard working and dedicated staff, but it also may have something to do with the chairman visiting the three operational sites on a regular basis to check everything is spick and span.

Our site visit starts from the main office block, workshop and coal processing area, which is still located on the Delhi section of the site, on the north side of Ponteland Road. The Brenkley site was designed to use the final void and the existing overburden tip on the Delhi site, in order to maximise the potential for progressive restoration. The challenge was getting overburden from the initial box cut on Brenkley, which lies on the south side of Ponteland Road, to the final void on Delhi and move material stockpiled from one site to the other, as the company continues to reinstate both Delhi and Brenkley Lane.

The solution Banks' engineers came up with was

A super-sized dust suppression unit in the shape of the converted Cat 777B model haul truck passes under Ponteland Road and the temporary Bailey bridge that allows access from the former Delhi site to the Brenkley Lane site. It will be removed during the final restoration phase in 2021.

to create an underpass large enough to enable two 91-tonne Cat 777s haul trucks to safely pass each other, with local traffic on Ponteland Road being carried above on a Bailey bridge. Banks used one of its massive 260-tonne RH120-C face shovels (from the Delhi site) to excavate the underpass beneath the Bailey bridge without disruption to the local community.

Banks' success in restoring surface mines, and working with the local community to minimise the impact of its operation, comes down to it starting comprehensive restoration work as soon as it has finished extracting the coal in a particular section of the site – known as progressive restoration. A great deal of time and care is also taken to design the soil storage mounds and overburden mounds so any visual impact of the operations is minimised. To this end, all mounds are sewn with indigenous grass seed as quickly as possible, so that they blend into the surrounding scenery.

Banks operates two rare Bucyrus RH120-Es – bought in the short period of Bucyrus ownership before Caterpillar bought out the entire Bucyrus product line of mining equipment in 2011 and both machines are looking very smart in Banks Mining's all-white livery, with

a Bucyrus purple stripe. The other monster miner on this site is an O&K RH120C.

I first stopped to see the Bucyrus RH120-E face shovel machine in action, loading lightly blasted mudstone into a modern fleet of haul trucks consisting of three Caterpillar 777Gs and four Komatsu 785-7s, along with some older, 91-tonne, 777D haul trucks. As part of Banks' ethos, the Bucyrus RH120-Es are fitted with the company's unique dust suppression system, whereby large 4,000-litre water tanks have been installed to the machines counterbalance and linked to spray bars on the top and bottom of the dipper arm.

This system is very effective at keeping fine dust under control when the bucket is either digging or when loading the trucks. The RH120-E water tanks are kept topped up by converted articulated dump trucks (ADTs) dust suppression units, with the tanks taking about 10 minutes to replenish during rest break periods. However, the converted ADTs' main duties are to dampen down the haul roads and the coal stockpiling areas. Banks also has a super-sized dust suppression unit in the shape of the converted Cat 777B model haul truck.

Converted articulated dump truck dust suppression units replenish the prime movers' water tanks during rest break periods; however, the converted ADTs' main duties are to dampen down the haul roads and the coal stockpiling areas.

The 120-E face shovel is five and half years old, with 13,000 hours on the clock, and its regular driver is Chris Richardson. Chris has been operating the machine for more than a year, having previously operated the older 120-C models, and is delighted with the extra levels of comfort and reduced noise found in the 120-E model. Chris said: "The 120-E is very responsive in the levers, has a lot of digging power, and produces fast loading times; not a bad machine to drive at all, it's a good piece of kit."

Banks uses both the Caterpillar Minestar and Komatsu Komtrax telematics systems to monitor machine performance, and Chris was happy to show me a hard copy of his cycle times operating the RH120-E; swinging a 15 cu.m bucket he can easily load the 91-tonne capacity haul trucks in three passes, in an average time of 80 seconds, and his best time was 75 seconds.

Over in the Durham Low Main section, a slightly younger five-year-old Bucyrus 120-E backhoe with 12,000 hours on the clock was sitting in the perfect loading position – on a 3m high bench – for loading blasted mudstone and sandstone into the back of the 91-tonne capacity trucks in the same fast cycle times

as the 120-E face shovel; that's good going considering the face shovel model has the legendary tri-power system to help its digging and loading performance. The backhoe machine was excavating the last few metres of overburden prior to hitting the coal seam and its performance has been enhanced with the fitment of a new 17 cu.m bucket made by local supplier MST.

This RH120-E backhoe was originally fitted with a 15 cu.m Bucyrus bucket, however Banks' senior management approved the investment in a bigger bucket in order to achieve a more consistent three pass loading with the Cat 777 and Komatsu 785 trucks. The new design from MST was made with Hensley ground engaging parts, consisting of a cast lip, shrouds, teeth and wing shrouds to protect the cheek plates. Bucket width is unchanged to retain high breakout forces, however, the design included an enlarged profile back wall to obtain an extra 2 cu.m of capacity and Banks are delighted with its performance.

To emphasise how mining shovels have evolved over the last two decades, Banks operates a comprehensively refurbished backhoe O&K 120-C version of this machine at its Rusha mine to undertake the same duties, which

Bucyrus RH120-E loading a fleet of Cat and Komatsu haul trucks with the cut moving in a westerly direction.

At the Durham Low Main section, a Bucyrus 120-E backhoe is sitting in the perfect loading position – on a 3m high bench – for loading blasted mudstone and sandstone into the back of the 91-tonne capacity trucks.

Looking due east, towards Blagdon Hall and Blagdon Park; this was once a void at the Delhi surface mine. This section has been restored with local grass, a large pond, with sheep and cattle grazing and is home to an abundance of wildlife, such as swans, hares and numerous smaller birds. In 2013, 51ha of the historic park at Blagdon Hall was restored and opened by Owen Paterson, then the Secretary of State for Environment, Food & Rural Affairs.

The oldest prime mover on site is an RH120-C model.

is to swing a 13 cu.m bucket. That is a big improvement in both product design and productivity. The team at Brenkley – who work a single 12-hour shift, Monday to Friday, due to planning restrictions – can move up to 140,000 cu.m of overburden and about 9,000 tonnes of coal per week, weather permitting!

Banks Mining has gone to considerable lengths to ensure its prime load and haul equipment meets its tough environmental planning conditions and, in addition to the spray bars, the two 120-Es have been fitted with large louvred grilles on the rear-mounted engine radiators and hydraulic cooling packs on the offside front. These modifications are to help suppress the noise from the large fan blades, and this technology is used extensively at other Banks' surface mines, both on the excavators and haul trucks where required.

Banks operates the latest haul trucks from both Caterpillar, in the shape of the 777G, and Komatsu's 785-7. The Cat 777G trucks are fitted with a Caterpillar-designed engine encapsulation package, a system Banks pioneered here in the UK many years ago, and is still in operation with its older Cat trucks across the model range. The Cat 777D trucks at Brenkley are fitted with full engine side panels, enlarged radiator cowling – with an acoustic splitter – and exhaust suppression kits, which all help to reduce the noise by a whopping 17dB(A). Put in context, a 10dB(A) reduction produces a perceived noise level drop of 50%. These modifications were designed and fitted by Banks to keep the noise within the site's environmental plan.

As mentioned in the previous chapter, the new Komatsu HD785-7 trucks are quieter in standard spec, however Banks has also fitted an extended radiator front cowling to allow louvres to be fitted to reduce the noise. This relatively simple modification leads to a significant noise reduction on the HD785s and we are told you can now hold a conversation standing in front of the radiator – at idle speed – not bad going considering there is a 1,200hp 30.4-litre V12 engine only a few feet away!

Banks Mining works hard at being a good neighbour, and this approach is not just limited to modifying the prime movers, this extends to drilling and blasting. Darren explained that this work is contracted out, but he is responsible for keeping this activity within agreed

A boost to Banks' environmental credentials, it is proud to be operating the first Tier 4 emission compliant Cat D9T in the UK – helping to save the environment and fuel.

limits. At the time of our visit, they were blasting close to Ponteland Road, and at this section of the site they only drill down to 4m and lightly blast to keep the noise and vibration down, leaving the big 1,500hp 120-Es to dig the material out of the face. Banks has set up monitoring stations around the site to measure noise, vibration and dust, and all this activity is recorded, and audited every three months by the local planning authority.

Another Cat machine helping to make an impact on the environment is the first Cat D9T Tier 4 machine in the UK. The very latest D9T dozer now features a Cat C18 ACERT engine that meets Tier 4 Final/EU Stage IV emissions standards and delivers 436 net hp at 1800rpm and a rise in torque of 36%. You may have noticed a slightly raised bonnet line; that is because sitting on top of the 18-litre engine are two big canisters comprising Caterpillar's Diesel Particulate Filter – which provides a particulate matter (soot) reduction of greater than 90% – and a Selective Catalytic Reduction (SCR) system that is claimed to provide a NOx reduction of greater than 90% through the use of AdBlue/urea solution, which is sprayed into the exhaust stream. The new D9T is

saving fuel too, as it is using 3 litres per hour less than a comparatively new non-AdBlue D9T on site.

The new D9T driver and site supervisor Alistair Renton was levelling a bench in preparation for an O&K 120-C face shovel to start work again. Alistair is a long-serving member of the Banks team and has been operating dozers for more than 30 years. He is very happy with the performance of the D9T: "I'm really pleased with this new dozer, as it's been trouble free, with a very modern and comfortable cab and has lots of power; it's the best dozer I've ever driven."

Over on the tip site another D9T was dealing with the constant stream of Cat 777Gs, 777Ds and Komatsu HD785-7 haul trucks. The skips on both the Cat and Komatsu trucks have been modified to Banks Mining's own standards and are fitted with enhanced load ejection on the tail and side spill boards for better load retention.

The coaling operation at Brenkley is handled by Volvo EC320s and Caterpillar 329Dcs coal shovels, which are fitted with front end equipment from Kocurek Excavators Ltd. The coal is loaded into the back of Cat

Cat and Volvo coal shovels are fitted with front end equipment designed for scraping thin seams of coal.

Photo overleaf courtesy of Banks Mining.

One of the new Cat 777G haul trucks heading to the tip site, with Komatsu HD785-7 trucks seen in the background.

Loading HGVs in just three passes are two Cat 980H wheel loaders; they are fitted with massive 9 cu.m coal buckets to efficiently move the high quality low density coal.

(740) and Volvo (A40E) adapted coal-bodied ADTs and transported to the processing area, where all coal produced is stockpiled, screened and blended before being despatched to the customer. Banks has permission to load and move up to 85 articulated trucks per day, and in order to keep the local roads clean they will pass through a wheel wash prior to leaving the site. Loading these HGVs are two Cat 980H wheel loaders, fitted with massive 9 cu.m coal buckets, which are able to load the HGVs in just three passes.

My tour finishes standing on top of the overburden pile looking out due west, with Brenkley Lane on the left (hence the name of the site) and a new bridleway made in the centre. On the left of the scene, new arable farmland has been created, along with small ponds to help encourage new wildlife. It is somewhat hard to believe this was once a large surface mine void on the Brenkley Lane site as the landscaping looks completely natural and untouched. The final restoration work at Brenkley Lane is planned to be completed during 2021.

Banks Mining has also set up a local community trust fund, whereby 10 pence is set aside for every tonne of coal extracted. This means a total fund of £240,000 will be available for grants if 2.4 million tonnes of coal is extracted from Brenkley Lane. Across its surface mine operations, Banks is keen to provide jobs for, and work with, the local community and businesses. Two examples of this approach are the fact the new 17 cu.m MST bucket was designed and made less than 20 miles away, and the order for a new D9T and 777G haul trucks was placed with the local Finning UK depot at Tyne and Wear.

Looking to the future, Banks has been working in partnership with Northumberland College to help identify local young people for apprentice positions at Brenkley Lane and its nearby Shotton surface mines. This is part of Banks' continuing commitment to providing direct benefits to the community through its local operations. As one of the biggest employers in Northumberland, Banks Mining supports skilled jobs for around 140 people at the nearby Shotton site, alongside a further 60 at Brenkley Lane, and it is estimated the two sites jointly contribute more than £35 million every year to the regional economy through wages, investment and the local supply chain.

Site map showing the former Delhi site restoration and Brenkley Lane operation plan, and aerial view of both sites during September 2015. Photos: Reproduced courtesy of Banks Mining.

CHAPTER 11

Banks Mining • Shotton surface coal mine • September 2015

As mentioned in the previous two chapters, Banks Mining currently has three operational surface coal mines: Rusha in central Scotland, Brenkley Lane and Shotton. The Shotton surface mine, Banks' largest, is situated at Cramlington, Northumberland, on the Blagdon Estate. The site covers around 342ha, which includes one of the biggest tourist attractions in the North-east, the world famous Northumberlandia – 'The Lady of the North'. It attracts more than 100,000 visitors per year to the 46 acre Northumberlandia Landform Park, which has free public access and 4 miles of footpaths on and around this large work of art.

Northumberlandia was privately funded by Banks and the Blagdon Estate at a cost of £3 million and used 1.5 million tonnes of material (predominantly compacted stone, clay and soil) taken from the surface mine and shaped by the construction and mining teams using the site's bulldozers and excavators. Visitors to the park, standing at more than 30m high at the head of this impressive structure, can also view Shotton mine in action. This is part of Banks' mining ongoing commitment to carefully restore its surface mines during current coaling operations.

There are approximately 6 million tonnes of coal to

Images taken from the viewing point at the top of Northumberlandia earth sculpture, from where visitors can view Shotton surface mine in action.

be recovered at Shotton and the site supports 150 jobs. Banks Mining has been extracting coal from this site since 2008, and has operated the largest hydraulic excavator in the UK, the 520-tonne monster miner Terex O&K RH200 face shovel, there since 2009.

Shotton staff achieved a new record in Banks Mining's 39-year history by producing 1 million tonnes of high quality coal from a single site in a single financial year in 2014. This was driven by particularly strong demand from its customers and this level of production could be achieved again – subject to market conditions. In April 2015, the Shotton team broke another important record, accumulating more than 250,000 hours without a reportable lost time work incident. Neil Cook, site manager at Shotton, puts these achievements down to having the support of experienced management colleagues, hard work and a team effort from everyone on site. Neil said: "We have a good health and safety record across the operation. That's because everyone from the plant team, operators and management work strictly to Banks Mining policies and procedures."

Banks continues to invest in health and safety training and has recently taken on a full-time trainer to help ensure the excavator operators and haul truck drivers are working as safely and as efficiently as possible. Banks Mining was delighted to report that in 2015 Neil had been awarded a Diploma in Quarry Technology by the Institute of Quarrying, after completing a three-year course run by the University of Derby. Neil has 35 years' service with Banks. Originally from Langley Park in County Durham, Neil joined the Banks Group in 1980 aged 20 and has since worked his way up through various roles to his present position, in which he's responsible for the operations and well-being of 150 people at Shotton.

Neil says: "Surface mining has changed a great deal since I first moved into the industry, and the environmental management standards to which we adhere ensure we always look to work to the highest environmental standards with sensitivity for surrounding communities, which is something that the diploma course has a very clear and specific focus on. And I've enjoyed Banks Mining's support in covering my travel, accommodation and studying costs, as well as the study time I needed to spend away from Shotton, which has been invaluable. Plus I've been able to feedback what

I've learnt to my colleagues and encourage new areas of best practice, which will be a central part of repaying the investment they've made in me."

Jamie Drysdale, deputy site manager, is a part of Neil's management team and my guide for the day. My tour starts at the newly relocated site offices and workshops, where Banks spent four months and invested more than £1 million to move the facilities in order to recover additional reserves of coal located at the former compound, known as Shotton south-west area. Banks Mining has an ongoing commitment to supporting local contractors, and they benefitted to the tune of £500,000 in supplying goods and services in supporting this project.

By UK standards, Shotton operates an incredible amount of prime movers and the largest haul trucks found on these shores (Cat 789B), with seven large excavators, comprising three RH120-C, two RH120-E, one Cat 6030 and, last but not least, a massive 520-tonne RH200 face shovel that develops 2,529hp from its two V12 Cummins K 1500-E twin turbo engines to swing its huge 26 cu.m bucket.

The current RH200 is such a well-respected machine that Neil and Jamie are both really looking forward to having two of these massive mining machines to help bring additional operational efficiencies to the site. The additional RH200 is a timely investment, as Banks Mining was granted a planning application to extract an extra 290,000 tonnes of coal at an area on the western side of the existing Shotton site, known as Shotton Triangle. The extraction will take two years from commencement to completion, and will run concurrently with the existing site. No overall extension of time is required for this additional extraction phase, with restoration of the entire Shotton site still scheduled to be carried out by October 2019. And with 21 million cu.m of backfill to move by 2019, Banks believes the additional RH200 will be a cost-effective investment and will help keep the schedule on track.

Banks Mining is also planning further ahead to keep this large fleet fully employed with a new application at a site further north at Highthorn, which will create at least 50 new skilled jobs as well as maintain the careers of at least 50 local staff on Banks' current operational sites. It will also provide an additional £48 million of economic contribution into the local supply chain, as well as many

The mighty RH200 working on a bench close to the famous Northumberlandia earth sculpture, seen at the top of the image.

Standing on a soil mound (near the A1 dual carriageway) looking due east, with Northumberlandia visible in the top centre of the image.

A new 320-tonne Cat 6030 swinging a bigger 16.5 cu.m Cat bucket (visor) to load a 136-tonne Cat 785 haul truck.

Drilling 7m down before blasting up to four times per day.

Komatsu D275ax (Cat D8 size machine) borrowed from the Rusha site, while Banks' own D9 was in the workshop for maintenance. Despite its size, it was coping well with a constant stream of 91- and 136-tonne dump trucks.

environmental improvements. In one example, if the Highthorn application is successful, Banks Mining has – as part of its strong 'development with care' ethos – negotiated the cessation of sand extraction along a mile-long stretch of the dunes on Druridge Bay beach, close to the proposed site.

Banks also continues to invest in new equipment and the latest multimillion pound prime movers. The latest 320-tonne Cat 6030 mining shovel was tearing out mudstone and sandstone with a newly fitted bucket visor from Caterpillar. The upgraded visor takes the bucket capacity up from 15 cu.m to 16.5 cu.m. The 6030 was quickly loading a steady stream of 136-tonne capacity, Cat 785 and 91-tonne Cat 777Fs and Komatsu HD785-7 haul trucks in five and three quick passes respectively. Working nearby are the drilling and blasting crew – which is contracted out – drilling 7m deep holes at this section for blasting later in the day. The depth of the drilling is dependent on the partings between the coal seams and also their proximity to the coal.

At the time of our visit Jamie Drysdale was also controlling the blasting operations from the highest point of the site, as this position provides an unrestricted view of the blast area danger zone and enables him to check

that all the sentries are in place. Access to the danger zone is prevented by positioning mobile plant across the haul roads at appropriate locations. However, before blasting can take place, water is sprayed all over the blast area to control dust during the explosion. And with seven hungry prime movers to feed, up to four blasts per day are carried out.

Banks has built numerous segregated roads around the site for light vehicle movements, and Jamie radioed in our position to the control cabin – situated at the top of the site – as we use this safe route to access the RH200 (Fleet number E163). During my visit, I spoke to Mark Guy, who has spent more than 15 years operating the RH200 face shovel, before being promoted recently to a site supervisor position.

Mark has operated every model of the RH120 series over the years and has no doubt the big RH200 is an outstanding piece of machinery above all others. Mark said: "The RH200 is such a comfortable and smooth machine to operate, the machine is completely planted with its 520-tonne weight, and is capable of digging some unblasted material without it rocking or being pushed away from the face."

Mark also said: "The full air suspension seat is a great place to spend a long shift and the hydraulic servo

Sentries posted to prevent access to the area during blasting operations. Banks has built numerous segregated roads around the site for light vehicle movements – as visible at the right-hand side of this image.

assisted leavers make the machine so smooth and a joy to operate; in fact, it's the best machine I have ever driven."

Mark and the RH200 started out at Pegswood surface mine, Morpeth, in 1995 and then moved to the nearby Delhi site before moving to Shotton. It may – subject to a successful planning application – move again to Highthorn.

The RH200 will easily load the 180-tonne capacity Cat 789 trucks in four good passes and the smaller Cat 785 trucks in three. Depending on material density the 26 cu.m bucket will lift 40–50 tonnes of material in one go, so the smaller 100-tonne trucks will not be used with it as they are simply not designed to have this amount of weight dropped into their skips in a single pass. It is also inefficient, due to the time lost spotting the truck under the big face shovel for just two passes. With twelve Cat 785 and two 789B trucks there are no plans at this time to increase the haul truck fleet to service the additional RH200.

Most of the prime movers work a 12-hour shift, from 7am to 7pm from Monday to Thursday with a 6pm finish on a Friday. However, the RH200 is such an efficient and important loading tool for the operation that it is double-shifted till 11pm, along with the two 180-tonne capacity Cat 798B dump trucks and sufficient number of Cat 785 trucks to keep the big excavator working at full tilt. Shotton currently enjoys an overburden ratio of 16:1, which means for every tonne of coal extracted the company needs to move 16 cu.m of other material. Neil and his team have a colossal overburden removal target of 309,000 cu.m (770,000 tonnes) to shift per week! This yields between 15–18,000 tonnes of high quality coal each week.

The RH200 was found digging 70m below the top of the mine surface to reach the Beaumont seam, with another 40m of extreme digging ahead to uncover four more coal seams. In fact, Neil explained that some of the sandstone is so hard that they will use the RH200 to lift

One of the biggest haul trucks in the UK, the 180-tonne Cat 789B is loaded and off to the tip site. The RH200 in the background is at a depth of 70m from the surface, with another 40m to go to reach four more coal seams.

Cat 785s teamed with the Cat 6030, and can be pass-matched to the RH200. Note the extended radiator cowling to suppress engine noise.

One of three converted Cat 777B on dust suppression duties on the haul roads.

the final section of overburden, and the coal seam will not be crushed by the weight of this huge excavator. The thickest coal seam at Shotton is the Northumberland low main (K100) at a healthy 1.6m. However, this was worked by underground miners in the past and has extensive old workings, while the Tilley stringer seam (P110) is the thinnest at just 150mm. Incidentally, coal seams tend to have local names that can vary across the coalfield, but are also identified by a series of alpha numeric codes that makes correlation from one area to another much simpler.

As part of Banks Mining's environmental management programme, the RH200 and other prime movers have dust suppression systems fitted. As covered in the previous chapter, this consists of large capacity water tanks fitted to the upper structure to feed spray bars on the stick – above and below the bucket – to keep fine dust in check. The water tanks are topped up from converted articulated dump trucks (ADTs) that now act as dust suppression units. Due to the size of the Shotton operation, Banks Mining operates three former Cat 777B haul trucks converted to dust suppression bowsers, and each truck can carry up to 40,000 litres of water in the back of their converted skips when full!

The Volvo EC320s' coaling shovels are a relatively new investment at Shotton, and a departure from the tried and tested Cat 329Dcs coal shovels. Both machines are fitted with front end equipment from Kocurek Excavators Ltd. The coal shovels load a fleet of both Cat (740) and Volvo (A40E) coal-bodied ADTs and transport the black gold to the coal processing area, where there is 135,000 tonnes of stockpiled, screened and blended coal waiting to be despatched to the customer. The adapted coal bodies (side extension) on the Volvo trucks are bolted on and this provides operational flexibility should coaling slow down – which is a rare occurrence – as the ADT bodies can then be quickly reconfigured for standard muck-shifting work, such as soil stripping.

Shotton also operates a large 85-tonne Cat 385B ME (mass excavator configuration) that has a number of tasks to perform, such as taking out the smaller partings between coal seams, as well as topsoil and subsoil muck-shifting duties. Neil reports that the Cat 385 is a very productive machine for this type of work.

Sitting on a massive progressive restoration area, about the size of 20 football pitches, is the smallest excavator on site, a 2.8-tonne Volvo EC25, used to dig test holes to allow compliance officers from the local council to check that top and subsoils are restored to the correct thickness (300mm topsoil, 700mm subsoil in this area).

In order to manage coal processing and stockpiling, Banks operates four Cat 980H wheel loaders fitted with high capacity 9 cu.m coal buckets, which are able

Cat D9T dozer and 16M grader working as a team to maintain the haul roads.

A Terex O&K RH 120-E loading overburden into a Cat 777F haul truck to reach additional reserves of coal located at the former compound known as Shotton south-west area.

Coal stripping, working across towards the highwall. Once coal extraction is finished, this area will become the haul road. Note the safety protection berm along the side of the Cat coal shovel.

A classic red RH120-C loading a Cat 785 next to a Terex O&K RH120-E loading a Cat 777D. Note the difference in the size of the truck skips – the 785 is a much easier target for the shovel driver to aim for during loading.

to load the HGVs in just three passes. The site has permission to load and move up to 190 articulated trucks per day. However, during its record breaking extraction (1,000,000 tonnes of coal in one year), the local authority granted Banks short-term temporary permission to increase this to 225 vehicle movements per day, as long as the trucks were not leaving the site in convoy.

Shotton's environmental plan also includes, among other conditions, that all the mobile plant is fitted with an automatic broadband noise reversing alarm. All the prime movers have acoustic insulation that reduces the noise on the CAT 777D from 94 to 78dB(A),

the RH120 from 90 to 80dB(A) and the RH200 from 97 to 82dB(A). This and other initiatives have helped Banks win two Noise Abatement Society awards. Noise reduction extends to its fleet of modern HGVs; Banks uses the latest waggons with at least Euro 5 engines for clean exhaust emissions, disc brakes – to enhance braking efficiency and minimise brake squeal – and rubber-bushed or air suspension to help reduce noise.

While Shotton operates a massive fleet of Caterpillar machines, which continue to perform well, it has also introduced Komatsu's HD785-7 91-tonne capacity haul trucks. These have been well received by everyone; drivers, because they are delighted with their comfortable

Some 135,000 tonnes of high quality coal is stockpiled, screened, blended and ready to ship to the customer.

In-cab shot of the RH200 loading a Cat 785 dump truck in three passes.

cabs and ride performance, and management as they are returning fuel consumption figures about 10 litres per hour less than the older engine technology found in the 777Ds. They are also highly reliable, so the maintenance team sing their praises, too. Derek Robson is the plant maintenance manager for Rusha, Brenkley and Shotton, and says of the Komatsu HD785-7s' performance: "The Komatsu trucks have about 13,000 hours on them and have proven to be very reliable since being built out of their shipping crates. We have not had to put a spanner on them outside of the normal programmed maintenance."

Derek was also complimentary about Cat equipment and service, as some of the Cat 785 trucks have more than 55,000 hours on the clock and are still going strong. Derek and his service team have built up a robust preventive maintenance plan, based on machine hours, total fuel burn, oil sampling, and operational experience, to swap out major components before failure occurs. Despite the age of some of these massive machines and trucks, 95% availability is not uncommon at Shotton.

With a second RH200 having been overhauled at Banks Mining's Thrislington main workshop, a short drive from Shotton, at the start of April 2016, the company has set another record as the only site in Europe to operate two 520-tonne Terex O&K RH200 mining shovels.

Shotton Progress Plan during August 2015.
Photo: Reproduced courtesy of Banks Mining.

CHAPTER 12

Coal Contractors Ltd • Demag H485 • Roughcastle opencast mine • 1986

Given that the largest hydraulic mining shovel in the world (in 1986) once worked at Falkirk and having been supplied with some great photographs of this massive miner by my good friend Nigel Rattray and also from the factory where it was built, I felt it would be remiss of me not to include details of the site and this ground-breaking machine in the mining chapter of this book.

The first Demag H485 built started work in September 1986 at Coal Contractors Ltd Roughcastle opencast mine Bonneybridge, in Falkirk, in central Scotland. The area of Roughcastle is near an ancient Roman fort and the Antonine Wall. The area can also trace its mining history back to when James Campbell & Co. (Roughcastle) Ltd, started coal mining around the late 1880s. As covered in Chapter 1, where there is coal you will normally find fireclay, and a few years later Campbell & Co. started to make firebricks from the clay material extracted along with coal. The materials were extracted by a number of drift mines to reach 10in thick seams of coal. After the Second World War more than 40 miners were employed and they produced nearly 100 tons of fireclay and 20 tons of coal per day!

Clearly digging coal and clay by hand, during the early years of underground mining, is a far cry from today's 'sunshine miners' using surface mining techniques, where the norm has become large mining shovels, dump trucks and 20- to 30-tonne hydraulic excavators adapted for coal shoving duties, matched to 35- to 40-tonne articulated dump trucks (ADTs), with adapted coal bodies to haul the coal to the stockpile areas.

Fast forward to 1986. Coal Contractors Ltd had operated two 180-tonne Demag H185 excavators, one face shovel and a backhoe model before taking delivery of the world's largest hydraulic mining excavator. This

landmark face shovel had an operating weight of around 540 tonnes and took four Demag engineers from the factory in Düsseldorf about two weeks to assemble the 16 main components, using two mobile cranes at the Roughcastle site.

The former Demag factory (now owned by Komatsu Mining Germany, KMG, as detailed in the Demag H255S chapter) is also credited with having built the first diesel-powered fully hydraulic excavator (a 12-tonne B504 model) in 1954, by fitting a hydraulic rotary valve to power the track motors while also maintaining full 360 degree rotation of the upper structure. In the space of just 32 short years of development the company has gone from the B504 model through machines such as the 180-tonne H185, 280-tonne H242 and 295-tonne H285 to reach a mining excavator, with the H485, that is 45 times heavier than the operational weight of its first fully hydraulic shovel!

As an aside, on visiting KMG's Düsseldorf based factory during 2012, and having received a presentation on the company's long history and landmark shovels produced over the years, there is a strong sense of legacy and pride in making some of the biggest and best hydraulic shovels in the world. The factory buildings have a presence of history about them as well, dating back to 1937 when a new excavator factory was planned and constructed shortly thereafter. Just inside the factory gates KMG also has a B504 face shovel on display that has been nicely restored in its then corporate blue colour.

As part of my visit, I was invited to lunch and was offered something to drink from a bottle clearly labelled Komatsu Hydraulic Oil! My hosts then explained it is a longstanding tradition for guests to drink the oil. Not wishing to offend my hosts, we toasted to good health and drank the clear liquid, which was of course some nice German schnapps — a great tradition and a very memorable experience for visitors.

Unfortunately I never got to see this impressive machine working, but having just returned from the Bauma 2016 exhibition, where KMG launched its new state of the art 691-tonne PC7000, I could imagine the H485 being close in stature to the new mining shovel, as the Demag H485's reach was only 0.7m shorter, from its front end equipment (at 20.3m) than the new Komatsu PC7000 at 21m. The driver's cab of the H485 was 8m off the ground, again very similar to the PC7000. However, that is where the comparisons end and, as you would expect, the modern PC7000 face shovel produces a much higher performance from its two 50-litre Komatsu engines. It is capable of a 5,000tph loading performance and produces a break-out force of 1.994kN fitted with a massive 36 cu.m bucket vs 1,800kN produced by the H485 with its smaller 23 cu.m bucket. That is still a pretty respectable performance given the 29-year age gap and the fact the H485 model is powered by one engine and has much older hydraulic technology.

Just like today's ultra-class rope and hydraulic mining shovels such as Komatsu's 800-tonne class PC8000, the cab on the Demag H485 is mounted on the right-hand side of the machine. This means that if the truck needs to be spotted on the left it leaves the shovel driver loading from his blind side, and anecdotal reports suggest that a number of the haul truck skips and canopies received minor impact damage as the 23 cu.m bucket occasionally made contact as it slewed around to load them.

The 37-tonne, 23 cu.m bucket had six massive teeth and five shrouds that were made by Bofors and secured by a mechanical locking system first patented in the 1960s by the same company. These ground-engaging parts were made from magnesium or chromium alloy and had to withstand the tear-out force of 183 tonnes, or 30 tonnes per tooth. That is equivalent to putting the weight of a loaded eight-wheel HGV tipper truck behind each tooth!

The H485 Demag was designed to load 160 to 200-tonne trucks at an average production rate of between 3,500 and 4,000 tonnes per hour depending on material

Coal Contracts' 540-tonne Demag H485 standing next to the first fully hydraulic excavator in the world, a 12-tonne Demag 504. Photo: Komatsu Mining Germany

density and ground conditions, and each bucket could hold approximately 50 tonnes of material. The H485 started out at Roughcastle loading four 136-tonne Cat 785 haul trucks, and then it was matched to load three large 180-tonne capacity Cat 789 trucks in three or four passes, which meant they could move the same amount of material with one fewer truck on a single shift system.

During the first two months of work at Roughcastle, some of Demag's research and development team stayed on to keep a close eye on H485 performance and then left two service engineers on site for a further 18 months, which was backed up by Volvo BM, the UK dealer at the time for Demag equipment, to provide aftersales support.

Unlike its German rival O&K, whose large mining machines were fitted with two engines, all Demag shovels in the range had one engine and in the case of the H485, a huge one at that – a 63.3-litre V16 cylinder turbocharged and intercooled diesel engine from MTU that produced 2,105hp at 1800rpm. The fuel consumption was designed to be around 250 litres per hour and at this rate the 6,500-litre fuel tanks could keep the machine operational for two 12-hour shifts without the need to refuel.

During a visit to another site, I spoke to Derek Taylor, who worked at Roughcastle as an O&K RH9 coal scraper operator, and he recalls a number of aspects about Roughcastle opencast mine around the time the Demag H485 arrived on site.

Coal contracts were working seven different seams in total; three of the coal seams near the top of the excavations belonged to British Coal and had a member of staff on site to manage the coal extraction. Derek recalls the site producing approximately 5,000 to 7,000 tonnes of coal per week, which equates to about 300,000 per annum. The coal was mainly transported to the power generating companies, such as the Longannet Power Station, Alloa – a short 30 mile haul – by railway waggons, as there is a narrow bridge at Tamfourhill, the most direct route to Alloa, which was unsuitable for goods vehicle traffic.

Prior to the Demag H485 arriving, the site operated

Planned maintenance work was carried out over the weekend. Note the engine top covers are lifted off and the side door is open to gain access to the walk-in engine bay. It was ahead of its time with a sloping rear mounted access step design to reach the operator's cab. Photo: Nigel Rattray

two Demag H185 excavators: one backhoe and the other a face shovel model. These would load 85-tonne Terex 3311 haul trucks. These two machines were replaced when the H485 become the primary loading tool, and while the H485 could carry out the work of the two H185 machines, if the H485 had a technical issue the site came to a halt! And this impacted most when the massive MTU engine was replaced twice during its time at Roughcastle. The site operated six days per week, from 7am until 6pm, Monday to Friday, and on Saturday morning, with planned maintenance carried out during Saturday afternoon and Sunday.

The coal seams at Roughcastle were thin, ranging from 2ft to 6in, however, the amount of material the Demag H485 could move to reveal them helped to make it a viable operation. The site operated two O&K RH9 coal shovels and two Komatsu PC130 excavators to scrape every last inch of coal. The buckets on the backhoe machine also had teeth welded to the back of the bucket to scrape away any remaining hard interburden on top of the coal seam. Two Komatsu 155 bulldozers were used for haul road maintenance, along with an

Aveling–Barford motor grader, and spreading material at the tip sites. A big Cat 245 excavator also worked at Roughcastle to lift some of the smaller coal partings.

The control levers on the Demag had a different operation pattern to today's machines. Derek recalls the levers working in an opposite layout to the standard pattern found on most other machines, with the boom and slew operated from the left-hand lever. However, drivers on the older RH9 machines were comfortable with the same operating pattern. Derek mentioned that the H485, with its 23 cu.m bucket, was restricted to loading the big Cat 789 trucks in three passes; otherwise the truck's hoist cylinders would struggle to tip its load of overburden material if a fourth pass was used. Derek also mentioned one of the truck drivers taking a Cat 785 to the site's weighbridge and found nearly 80 tonnes loaded from just one bucketful! This would be at the extreme end of the H485 lifting capabilities and a fully heaped bucket of highly dense material. Clearly the company needed a bigger truck to match the H485, hence it ordered the three Cat 789 haul trucks.

Derek also recalls that it was a great place to work

Roughcastle was a large 220-acre site and in the background the Demag H485 is loading a haul truck on its blind side while a Komatsu PC130 excavator and small O&K RH9 face shovel are busy recovering the valuable coal. Photo: Nigel Rattray

Image overleaf courtesy of KMG.

IMAGE COURTESY OF NIGEL RATTRAY

The Demag H485 has just finished loading one of the Cat 785 haul trucks as the other one arrives for the next load. Note the small Komatsu 155 dozer tidying up the loading area. Photo: Nigel Rattray

and there was a good feeling of camaraderie among all the operators, along with high attendance rates and low staff turnover, with the Demag H485 only operated by two different drivers during its time at Roughcastle.

Roughcastle opencast mine was worked until 1996 and the 220-acre site was then reinstated with the help of a new Demag H135S face shovel. The reinstatement plan provided one third farmland and two-thirds to create a community woodland area, which has a network of footpaths, picnic spots and viewpoints. Coal Contractors also paid for tree planting, which comprised native species found throughout Scotland.

I understand that the Demag H485 remained at Roughcastle until coaling operations finished and was then sold, disassembled and shipped to Canada, where it is presumed to have been scrapped after a long and productive life.

Thanks to Komatsu Mining Germany for supplying some images for this chapter, particularly the one showing the Demag H485 loading a Cat 785 truck in the first few days of operation as it is a rare shot of the Demag H485 loading from the shovel driver's side, but the blind side for the Cat 785 haul truck driver.

A mining band of the Ruhr area of Germany standing in the bucket of a Demag H485, this time with a 26 cu.m capacity. Photo: Komatsu Mining Germany

Quarries

One of my first assignments as a photojournalist was to visit Caterpillar's renowned D9 track-type tractor in a silica sand quarry. The D9T model replaced a long list of D9 models used as a production bulldozer to extract silica sand for glass making. Shortly after that, I visited a young lady driving the first Terex TA400 articulated dump truck sold in the UK; to Garriock Bros in a hardstone quarry in Edinburgh. Her name was Tina Neil, and she demonstrated a quite exceptional level of knowledge of the truck and skill operating the ADT in a compact area.

Lafarge, at its Dunbar cement works, made a significant investment in two Komatsu PC2000 excavators and a fleet of Cat 777F dump trucks during 2012 to 2013. I was able to capture not only their operational performance, but also the build process of the first PC2000 face shovel model in Europe.

I also made a number of visits to Wm Thompson & Son, at its Dumbarton quarry near Glasgow, to see the first Cat 336E Hybrid excavator sold to a UK operator, which promised to bring significant fuel savings over a five-year period of operation. The company also became the first to buy Caterpillar's 972M XE wheel loader model, fitted with a constantly variable transmission, which would also contribute to a reduced fuel bill for this site.

Over in the north-east of Scotland, Jim Jamieson Quarries Ltd invested in the UK's first Liebherr R960SME excavator for its Ardlethen Quarry just north of Aberdeen. There it would not only be used to load the primary crushers, but would also eliminate the need to drill and blast some sections on the quarry by fitting a massive ripper tooth attachment for rip and load duties.

Two things were obvious during these visits: one was how much the operators enjoyed driving the latest technology; and the second, was the high level of service operators receive from their main dealers. This plays a significant part in the decision-making process when purchasing their next machine.

One of the most interesting visits was catching a boat to Aggregate Industries Yeoman Glensanda super quarry on the west coast of Scotland, where CA Blackwell (Contracts) Ltd was awarded a four-year load and haul contract for the site. It shifted approximately 7 million tonnes of blasted rock each year from the quarry face to the processing plant using eight Cat 777G haul trucks and a pair of massive 100-tonne 992K shovels – Caterpillar's biggest wheel loader in the UK. And since access to this quarry is by sea only, the massive production output makes Glensanda one of the biggest ports by tonnage in the UK.

CHAPTER 13

Geddes Group • Caterpillar D9T Devilla Quarry • June 2010

Geddes Group is a family-run business, managed by three brothers, with its head office in Arbroath on the north-east coast of Scotland. It has continued its long standing relationship with Caterpillar by investing in a Cat D9T bulldozer for heavy duty production ripping and dozing duties in Devilla quarry, a silica sand pit owned by O-I Manufacturing (UK) Ltd (formerly United Glass) in Alloa, central Scotland.

The brothers' father started out with an agriculture business in 1939 and bought his first Caterpillar in 1947, a 48hp D4 track-type tractor fitted with a ploughing attachment to service the local farming community. A scraper box was soon purchased to further use the D4's capabilities. Geddes has business interests across some very diverse sectors, including waste management, public works contracting, ready-mix concrete production, skip waste disposal, recycling operation, plant hire, and sand and gravel quarries, and it still owns and operate farms, as Geddes Farms.

To say Geddes Group has a long-term business relationship with Caterpillar is a bit of an understatement. With more than 60 years of Cat equipment experience under its belts, this partnership has helped the Geddes Group grow a strong and diverse business, with some 60-plus pieces of Cat equipment operating within the group. Geddes has been operating in quarries since the early 1970s and has been active in both hard rock and sand and gravel workings. During this time, it has operated in seven sand and gravel pits, and provided screening, crushing and washing operations to supply a steady stream of sands and gravels for customers in the Tayside, Fife, Perthshire and Edinburgh areas.

The silica sand operation at Devilla quarry is situated just off the A904, close to Kincardine Bridge. The quarry started operations in 1966 and is a large 600-acre site, rich in high quality silica sand. The site enjoys only a very thin layer of overburden material and has a 20m-deep seam of sand to extract, allowing the quarry to produce around 300,000 tonnes of material per year. The silica sand is transformed into glass bottles for a number of famous whisky and gin brands at the nearby O-I bottle manufacturing plant, which produces more than half the amount of glass bottles needed to service the Scottish spirits industry.

The D9T dozer is an essential part of the quarry operation, as any loss of production at the rock face quickly means shortages downstream at the glass factory, with serious production and economic consequences for all concerned. However, having said that, the D9T has some production capacity in reserve, which with good forward planning has allowed it to be off site for a few days to help construct access roads at a local wind farm project and the like.

Geddes Group has held the contract for supplying a production bulldozer at the Devilla quarry for more than 15 years and during that time it has operated a number of second-hand, low hours, dozers, starting with the Cat D9G model. The conventional low final drive sprocket on the D9G model was replaced by Cat's revolutionary elevated final drive sprocket model, found on the mighty D9N; it was followed by the successful D9R model, which was finally replaced by investing in a brand new £450,000, 48-tonne Caterpillar D9T with ripper attachment.

Here in the UK, heavyweight bulldozers – such as the Cat D9 series – are mainly found taking care of surface coal mine haul roads, or involved in levelling duties at the mine's tip site. You may even find one or two in some large road construction projects. This Cat D9T has a number of interesting modifications to help it cope

Devilla quarry is a large 600-acre site, rich in high quality silica sand deposits.

Devilla has its own equipment, a Volvo L150E wheel loader to load a fleet of Volvo A30D articulated dump trucks.

The sand and rock is highly abrasive, as can been seen with plumes of dust coming off the blade's cutting edges as it strikes against the hard rocks.

in this unusual application. Silica sand is a very abrasive material, which is particularly challenging for a dozer's engine air intake system, undercarriage and tracks. Notwithstanding Geddes Group's preference for Cat equipment; in this application the D9's elevated sprocket design is ideally suited to the harsh environment, by keeping the final drive train components and seals away from the fine sand particles.

Bruce Geddes, engineering manager, explained that experience with the D9G model, operating a low final drive sprocket dozer, in this application, can cause a problem in keeping the final drive units and seals serviceable. Bruce was very impressed with the performance of the D9R with elevated sprocket, as the machine had no final drive issues during its operational life, despite the fact it was operated for many years in this application with only one set of drive bearings fitted as part of planned maintenance. Furthermore, to build in more durability for the D9T dozer, it was specified with Cat's super extreme track grousers with Abrasion Resistant Material (ARM) shoes, which should significantly reduce unbalanced track shoe wear and deal with the conditions better than a standard track pad.

Geddes plant engineers were ahead of the US giant's design team – long before a track carrier roller first became an option on the D9T model – by fitting a carrier roller to their D9N dozer. This modification was designed to address excessive wear on the undercarriage and especially on the reverse side of the track. The wear issue was pinpointed to where the track sagged between the idler and the drive sprocket. Having fitted a carrier roller, this gave them an extra 1,000 hours life from the components, a small but very effective design change. This modification was so successful Geddes workshop staff changed the carrier roller over from the D9N to the D9R and then took the Cat option on the D9T model.

The D9T bulldozer is powered by an 18-litre, C18 ACERT Tier 3 compliant engine, producing 410hp at 1800rpm. It is matched to a high efficiency torque converter and power shift transmission with three forward and reverse gears. The C18 engine has an air-to-air intercooler and water-cooled turbocharger for sustained high power output, with the turbo breathing through an optional special turbine air pre-cleaner stack.

This Cat option had been specifically designed to deal

with very dusty and abrasive material applications, as found at Devilla quarry. The turbine provides improved engine air filtration by using the optimax dual-stage pre-cleaner powered by the engine's intake and exhaust airflows. Intake air is spun by a flow-driven impeller, debris stratifies along the outer wall and is then ejected back into the environment. Remaining contaminants are collected and removed by a secondary scavenger system, allowing only pre-cleaned air to continue to the engine's air filter element.

With 410hp and 49 tonnes to play with, the driver controls speed, direction, and steering via an electro-hydraulic single tiller control lever that operates Caterpillar's differential steering system (diff-steer). It is used to its full potential in this application, as the driver spins the dozer through 90 degrees numerous times to cross rip the material and to get the single shank ripper in close and tight to the pit wall that is quickly formed by this big powerful dozer.

The single shank ripper is operated via a rigidly mounted handgrip control, which provides support when working in this tough terrain. The low effort thumb lever controls are used for raising and lowering the shank-in and shank-out positioning, with a thumb button automatically raising the ripper. The ripper shank depth is adjustable via a pin puller mechanism that is controlled from the cab. The driver said he likes the ease of controls, particularly in this application given the large number of ripping cycles made in comparison to the dozing effort.

Additional driver aids include a rear-view camera and LCD screen fitted by Spillard to enhance the standard rear-view mirrors and provide a safer view to the rear when reversing and ripping. The MSP autolube system is an option the Caterpillar dealer, Finning UK, has fitted to cover the 16 lube points and to help improve the life of the ripper linkage components. It also makes regular maintenance duties easier for the driver.

Once the D9T dozer has built up a sufficient stockpile of silica sand, the quarry's own Volvo L150E wheel loader is used to load a fleet of Volvo A30D articulated dump trucks to haul the sand a short distance to the processing plant, where the primary crusher produces rocks about 4in in size. It is then screened with the lower quality sand being used at golf courses, etc. Devilla quarry produces a high ratio of good quality silica sand,

some 65 tonnes for every 80 tonnes excavated. Once the sand has been fully processed, it is then transported by road, with the vast majority being delivered to the nearby I-O glass factory.

This D9T is fitted with a standard 13.5 cu.m semi-universal blade and also has standard – bolt on – cutting edges. These DH-2 steel cutting edges are expected to wear at normal rates, however, Bruce intends to upgrade the edges to an abrasion resistance set for extra durability. Geddes group has its own in-house team of maintenance staff to carry out 99% of the regular maintenance duties and major overhauls at its Arbroath workshop. The first major overhaul for the D9T is expected to be at around 20,000 hours, and Geddes group will expect another 20,000 hours out of the machine.

Frank Geddes, MD, believes loyalty to the Cat product is built around the firm's philosophy of building equipment that can be rebuilt. Frank said: "It does not matter how old a Cat machine is, you can always get a spare part for it and if we place a parts order with Finning before 5pm they can be trusted to deliver it next

The D9T cross rips the material to help with the dozing phase of the operation.

One of a fleet of Volvo A30D articulated dump trucks to haul sand a short distance to the processing plant.

Cat D9T parked during a rest break, with the processing plant in the background.

Geddes obtains 1,000 hours more life from the track chains by fitting a carrier roller first fitted on its D9N and D9R. It then took the Cat option on the D9T model.

day to our workshop by 2am." Bruce also has nothing but praise for Finning's parts service. He is able to log on, locate and assess parts availability throughout the entire Finning network of depots in a matter of minutes, then direct any of his six fitters to obtain a part from the nearest branch to keep the machines down time to a minimum.

One task that the fitters carry out on a regular basis is the replacement of the D9T's ripper tip, a task that is carried out about every 10 to 14 days due to the considerable amount of ripping cycles performed in this application.

Geddes Group took the option to fit the D9T with Caterpillar's product link. This option allows Finning, Bruce Geddes and his workshop staff to monitor machine performance, diagnostics and fuel usage. One of the great benefits of using product link data is the ability to get early warning of fault codes and repair before failure.

Sitting in a comfortable seat, this is the driver's view to the front of the 13.5 cu.m semi-universal blade.

The driver takes an aggressive angle of attack to fill the blade as efficiently as possible.

The driver of the D9T has operated all other pre-owned D9 models used on this site and is happy that he has a shiny new dozer for the very first time. He particularly likes the comfortable driver's seat and spacious cab, which Caterpillar boasts is 15% wider, enjoys 30% more glass area, and has a significant 50% reduction in sound levels over the previous D9R dozer cab. Caterpillar's diff-steer is a hit too, as the D9R only had the older and less effective clutch mechanism design with which to steer.

Geddes Group is confident the Cat dozer will provide it with many reliable and productive hours and serve it well into the future.

CHAPTER 14

Garriock Bros Ltd • Terex TA400 Articulated Truck and Tina Neil • June 2010

Terex dealer TDL sold the first TA400 articulated dump truck (ADT) to Garriock Bros Ltd, and it was driven by a young lady named Tina Neil at a quarry operated by Cemex at Newbridge, Edinburgh. I was interested in what Tina had to say about the top of the range Terex ADT.

Garriock Bros Ltd is a family owned company that started trading in 1975. With its headquarters in Lerwick, Shetland Isles, the company employs some 180 people across six divisions. In 2005 it became the official distributor of Metso Minerals Construction and Recycling Equipment in Scotland, providing many major quarry operators with Metso equipment and plant hire services.

Michael Broe, regional director (Edinburgh), is in charge of the Edinburgh operation that took over an existing crushing contract in Bonnington Mains quarry in 2002. Since then it has made significant investments in

Shot taken from the haul road looking into the pit. Note the tip site/primary crusher on the left-hand side of the image. Garriock Bros Ltd made significant investments in new Metso equipment at Bonnington Mains to help produce 250,000 tonnes of material per annum.

new equipment and is firmly established as a key player in the aggregate and heavy plant hire business.

The Terex TA400 is a part of that investment, and the main haul truck at the quarry. The Bonnington Mains aggregates quarry, situated just off the M8 motorway at Newbridge, is capable of producing some 250,000 tonnes of high quality basalt stone per annum, with the TA400 used to haul the blasted material to the tip site next to the primary crusher and stockpiling areas.

Tina's career started in the construction industry in 2006 after leaving university with a degree in HR and Business Management. However, Tina decided that an office job was not for her, and with her love of cars and karting. Tina's husband (a 360 excavator operator) suggested getting an ADT 'ticket', which she funded out of her own pocket and she has not looked back since.

Her first job was driving a Volvo A30D ADT for a local plant hire and tipper company in her home town of Denny, where she gained experience on a number of muck-away sites across central Scotland over a two-year period, before joining Garriock Bros.

The Cemex site was not her first quarry experience as an ADT driver, as she was assigned to another Volvo A30, this time at the Craigie quarry in Kilmarnock, South Ayrshire. Tina was then asked to work in the Edinburgh quarry driving a Terex TA40, the predecessor to the TA400. Her stint on the TA40 lasted for about two years, during which time she gained a lot of experience operating the big Terex truck. She got her hands on the TA400 shortly after the dump truck was first displayed at the ScotPlant 2010 exhibition, which is very close to the quarry site in Edinburgh.

Tina is not only an experienced ADT driver, but also holds a licence for a wheeled loader. When asked about the new truck's performance, Tina clearly has a good depth of knowledge on both the TA400 truck and previous machine as she gave me a very analytical and articulate description of the pros and cons of each's abilities on different terrains, with Volvo getting her vote for muck-shifting and the TA400 getting the big thumbs up for rock solid performance in quarrying applications.

It is fair to say Tina is impressed with the performance of the TA400, as it has very good traction, a sentiment backed by Terex's ADT global product manager. He describes it as the best in class for rimpull performance, with the Detroit diesel series 60 engine producing 2,100Nm maximum torque at 1350rpm. Tina also thought the cab interior and driver comforts were far superior to any previous ADT she had driven, in particular the driver's seat, which is a vast improvement over the previous Terex TA40 series truck, as it has air suspension and arm and headrests fitted.

With the help of the mirror and rear-view camera package, Tina handles the reverse along the narrow haul road with ease.

The TA400 offloading at the tip site. Note Tina's name is written on the cab side panels.

Tina carrying out oil and water level checks, made easy via the power-lifting engine hood.

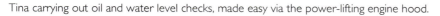

Tina also appreciated the small design details, such as the fact the cab interior trim is now a mix of grey and matt black, as the previous mono grey trim on the TA40 was too much like the quarry stone in colour and gave no contrast between the cab interior and the outside world. The product development team at the Motherwell factory redesigned the layout of all the main control switches; for example, the switches that set the engine and transmission retarders are nicely laid out and within easy reach, with the gearshift selector also well placed on the right-hand side to control the Allison six-speed fully automatic transmission with six stage retarder linked, via an electronic control module, to the engine Jacobs (Jake) brake.

By her own admission, Tina is not the tallest person in the world, and appreciates the other small, but important, design details Terex has put into this truck, such as the easy to reach drop down grease lines, as illustrated at the body tailgate main pivot pin, and items such as ground level sight glasses for hydraulic oil levels and the easy to use electrically powered engine hood, which lifts to check and top up engine oil and water levels as part of the driver's daily routine.

Once Tina had given me the guided tour in and around 'her truck' (well Tina's name is written in big red letters on the cab side panels) it was time to buckle up for a few runs to the tip site. A big 45-tonne Hitachi Zaxis 470 LCH backhoe excavator began loading us with its 3 cu.m rock bucket in six passes and made sure we were fully loaded with 38 tonnes of rock. With Tina in the driving seat, I settled into the small passenger seat and found a good spacious cab for two people, with all the usual refinements such as air con, radio and CD player – it even has an MP3 socket for your iPod. In fact, the TA400 cab layout and dashboard materials are just what you would expect to find in a modern car.

Once we got rolling, with the big Detroit diesel engine pulling cleanly away, it was clear that Tina's air suspension seat was dealing with the haul road much better than my unsprung passenger's seat, and she made the point that for driver comfort and the ability to sustain a 12-hour shift in the TA400, the driver's seat in this truck is a 100% improvement on the previous model and helps to make her a more comfortable and productive driver.

Other features that makes the truck easier to drive is the standard fitment of a reversing camera, with its bright large LCD screen mounted on the top right-hand corner, giving very good rearward visibility, along with four effective rear-view mirrors. They both work well in this application as the manoeuvring area at the bench is not that large, resulting in Tina sometimes reversing the length of the top part of the haul road in order to avoid making numerous shunts at the spotting area, so the camera proved to be a useful tool for spotting the TA400 under the waiting Hitachi 470.

Other impressive features noted during the run were the adjustable three-position automatic engine 'Jake' brake fitted to the 14-litre, six-cylinder, 450hp Detroit diesel series 60 engine, which can be adjusted to suit the terrain. I noticed as soon as the engine was above 1350rpm on the overrun the engine brake and gearbox retarders would cut in, providing progressive and effective retardation throughout the descent. This reduces the need to apply the foot brake during the haul cycle, which will translate into lower running costs and down time. Tina did comment that the oil-cooled wet disc brakes work extremely well when you need to stop quickly. This is a very clever truck, a bit like its driver, as it changes gear, brakes and can play your iTunes list automatically!

Tina looked totally relaxed as she continued to manoeuvre the big 3.63m wide truck down the narrow haul road, as the management team at Bonnington quarry uses the maximum amount of available floor space, with a large number of mobile primary and secondary crushers, screens and stockpiles to negotiate at the bottom of the pit. Tina continued to demonstrate her driving skills by making a sharp left-hand turn at the bottom of the haul road and then proceeded to make her way through a tight gap between the water treatment pond and a stockpile of newly crushed rock, before reaching the primary crusher and tip site area. Once there, the TA400's electronically controlled, double acting hoist cylinders bumped its load of basalt rock through the hinged tailgate in just 13 seconds and on to the stockpile area for the Hitachi W310 wheeled loader that will feed the Metso primary crusher.

After my temporary chauffeur – Tina – dropped me off, I made the short ½ mile journey from the quarry to Garriock's plant hire and workshop facilities to catch up with Michael Broe and find out what he thinks of the TA400.

The TA400's car-like interior with all the main control switches nicely positioned on the dashboard. Note the narrow haul road exit ahead.

Tina skilfully negotiates the big TA400 between the stockpiles and the water treatment pond.

The TA400 making its way back from the quarry floor up the twisty haul road.

With Michael having had a good experience and low running costs with the Terex TA40, he was very impressed with what he seen and heard at the TA400 launch day, held at the Motherwell factory, and started discussing a deal to replace the existing truck with his longstanding local TDL account manager, Hamish Ross.

It's the old adage in truck sales, sell the first truck and the service back-up will sell the second truck for you, as Michael explains: "We enjoy great service from Terex, with its industry leading 24-hour parts delivery or the parts free guarantee. When you couple this to a factory that is a mere 30 minutes' drive away, I believe Terex service back-up is second to none." With a large modern workshop and experienced fitters, regular servicing is carried out at 500-hour intervals by the in-house team.

Michael went on to explain how the TA400 is performing: "Apart from a very minor problem with some loose bolts on the back door the TA400 has been faultless, providing very good fuel economy in line with the manufacturer's claims." He feels the design approach of a wider and low centre of gravity body is the right way

to go, resulting in a full payload that does not need to be heaped, meaning less spillage around the site.

Michael's final word on the TA400: "Terex put a very attractive package together, and I'd buy another one tomorrow."

Michael is very impressed with Tina's performance too, as she just gets on with the job, looks after the truck really well and, dare I say it, in the predominantly male-oriented world of construction and quarrying, she has earned the respect of the rest of the quarry team because of her ability to handle the big Terex truck with ease. Michael has also invested in staff training, including Tina's, by funding a Minerals Product Qualification Council MPQC course, and Tina has achieved a level 2 certificate in operating plant and machinery.

At the time of writing this book, Tina had left the quarry when the crushing contract finished and had taken a sales manager's job with Hodge Plant Ltd, the main agent for CASE Construction Equipment in Scotland. Her years at the quarry face have served her well in this new role.

At the 2014 ScotPlant show, where Tina has a sales manager's job with Hodge Plant Ltd, the main agent for CASE Construction Equipment in Scotland.

CHAPTER 15

Malcolm Construction • Terex TR45 rigid truck • Loanhead Quarry • November 2010

Malcolm Construction, part of the Malcolm Group, has chosen a Terex TR45 rigid quarry truck, to operate at its Loanhead quarry situated off the A737, Beith to Lochwinnoch Road, in North Ayrshire.

Malcolm's Loanhead quarry is 25ha in size with reserves for approximately 25 years, and is expected to produce, at current production rates, 500,000 tons of basalt rock per annum to supply all Malcolm Construction projects, such as new schools and a national customer base throughout the UK.

Loanhead quarry has been in operation since the late 1960s and has changed hands many times, with Malcolm Construction acquiring and operating it since 2007. During this time, with a large fleet of HGV tippers at the group's disposal and augmented by some Terex articulated dump trucks (ADT), both these types were used to move crushed rock from the elevated benches down to the quarry floor. However, this arrangement proved to be a less than ideal situation and, after a long evaluation, the company decided to invest in a 41-tonne

The quarry is 25ha in area with current reserves for approximately 25 years, and is expected to produce 500,000 tons of basalt rock per annum. Malcolm tipper trucks can be seen being loaded with finished products.

capacity Terex TR45 rigid haul truck for all hauling duties, in order to save costs and double the payload per haul cycle.

The TR45 is not the only Terex equipment on site, as Malcolm Construction also operates Terex mobile Pegson jaw and cone crushers to process the quarried whinstone, where they will generate a range of eight specified quality materials including Type 1 Sub-Base CL803, 0/80mm crusher run, 40/80mm scalpings, 40mm to 10mm shingle size and 0/6mm crushed fine aggregate.

David Balmer, operations and environmental manager, explains the buying decision: "Having recognised the need to control costs and improve efficiency we started to look at the option of a quarry truck. We have a long-standing relationship with the Terex Group and its products, from mini diggers, ADTs and Pegson crushing equipment. While we looked at competitor trucks, the Terex TR45 was our first choice."

David also comments that: "We enjoy good after sales service from Terex, which helped clinch the deal as the Terex factory at Motherwell is only an hour's drive from the quarry."

Some of the key modifications to the TR45 were carried out at the factory; Malcolm Group would normally paint all its equipment in-house at its purpose-built MOT test centre and bodyshop at Linwood on the south side of Glasgow. However, since the TR45 was just a bit bigger than it could accommodate, the company asked Terex to paint the TR45 at its factory. Terex's product manager explained that this paint job is not just a simple spray-over on a standard production white truck, as Malcolm's TR45 was painted from scratch and from top to toe, including a black gloss chassis, in its very distinctive Malcolm Construction livery. Terex is happy to provide other customers with a bespoke livery.

The dump body skip sides were also modified during the build process; the height was lowered by 150mm in order to make the truck a better match for the two Komatsu WA500 wheeled loaders to quickly four pass load the truck. Three optional extras were specified: an automatic chassis lubrication system to help reduce maintenance costs; an amber warning beacon, now standard throughout the Malcolm Group; and a 'white noise' reversing alarm to help keep the noise inside the quarry. As part of the package, Terex trainers provided

operational training to a number of drivers who have come from an HGV background.

Once the evaluation process and specification details had been completed, the Malcolm Group's chief executive, Andrew Malcolm, finalised the deal with his longstanding local contact, Terex sales manager Hamish Ross. Mr Malcolm said: "We have a long standing, working relationship with TDL and they are always responsive to our requirements. Our decision to invest in a Terex machine was made easy by our knowledge and confidence in the excellent after sales service provided by them, which in our experience is second to none. The truck is good quality and we look forward to working with them again in the future."

Bob Fulton, Loanhead quarry manager, also helped to select the truck; praising the TR45 heated dump body – as the hot exhaust gases exit through the body skip – as this will be a useful feature during the winter months when they are hauling wet material or during periods of overburden removal.

Bob explained that the TR45 has been well received by the drivers and has been primarily assigned to Bobby Gilmour, an experienced HGV driver. However, after a short time behind the wheel, he is starting to get comfortable handling a truck of this size. The TR45 is more than 4m wide; (HGVs are 2.5m wide) while the length and height are not that dissimilar to an eight-wheel tipper truck, at 4.5m tall and 8.7m long.

To get a first impression of the TR45 from the very spacious cab, as used for the TR100, (91-tonne flagship of the rigid range) I took advantage of the truck's instructor seat. The cab is equipped with the usual refinements found in a modern quarry truck, such as radio and CD player, air con, comfortable air suspension seat with four-point harness, large full width front sunscreen and all-round tinted glass. Bobby comments that he is really pleased with the truck, as it has significantly more power and a better braking system than the eight-wheeler counterparts, leading to improved cycle times.

During my time in the cab I observed Bobby making good use of the hydrodynamic retarder. As we headed for the tip site below he constantly modulated its use via the large control lever, adjacent to the gear selector, on the right-hand side of the cab. The powerful 700hp retarder is fitted to the six forward and two reverse speed Allison transmission and was providing nearly all the braking effort during the long descent, with the rear oil-cooled and front disc brakes only coming into play at the steepest part of the haul road as the truck neared the quarry floor. This useful feature will help to keep down time to a minimum and reduce maintenance costs.

The Malcolm group has its own team of experienced mechanics to maintain the truck, however, during the warranty period Terex will carry out regular servicing at 500-hour intervals for the engine, and the transmission at 1,000-hour intervals.

The TR45 is equipped with a reversing camera, with its bright large LCD screen mounted on the lower right-hand corner of the dash, giving very good rearward visibility. With four effective rear-view mirrors and two close proximity mirrors that give a good view to the front and rear of the truck, the rearward visibility package works well in this application due to the high volume of site traffic, with both wheeled loaders and HGVs working around the quarry floor area.

The TR45 has a 26 cu.m capacity body, with a longitudinal 'V' shaped floor that is made from Hardox steel and rests on impact absorption pads to help dampen the shock during loading. The skip has a good host time of around 13 seconds to discharge the full 41-tonne payload, with superior stability during tipping. The skip lowering times are much better than a standard HGV tipper. The twin inboard host cylinders have a two stage construction, with a power down function coming into play during the second stage of operation, which helps to lower the body in just 9 seconds.

Terex manufactures its own strut-type independent front wheel suspension units, using self-contained, variable rate, nitrogen/oil cylinders. The rear set-up has variable rate nitrogen/oil cylinders too, with A-frame linkage and lateral stabiliser bar.

Terex claims the large 21.00 diameter wheels and tyres offer a performance (better ride and less tyre wear) advantage over competitor machines in this class size fitted with 18.00 tyres, but the claim could not be tested on the day as the haul road is in very good condition, maintained by the two Komatsu WA500 wheeled loaders and a pair of skilled operators.

Under the classic bull-nose bodywork, sits a powerful

Easy access to the engine oil dipstick, as part of the driver's daily checks.

The TR45 seen here coming down the super smooth haul roads.

The retarder (brake) lamp on during the descent to the quarry floor.

18.9-litre Cummins QSK19-C525 diesel engine, fitted with electronic controlled direct fuel injection, and an air-to-air charge-cooled turbocharger. The engine produces 525hp at 2000rpm and peak torque is 2,407Nm at 1400rpm. Clearly the engine produces plenty of low down power and meets Tier 3 and EU non-road mobile machinery directive Stage III emissions standards.

At the time of my visit, crushing equipment is situated at the face, where a JCB JS360 excavator loads Terex mobile Pegson jaw and cone crushers from an elevated position, with the two crushers working in tandem to produce specified quality materials. The material is loaded into the TR45's skip by the big Komatsu loader, and then hauled to the quarry floor where the load is discharged into a stockpile area for the other WA500 to feed the screening equipment. The quarry manager is planning to change this arrangement by moving the two crushers down to the quarry floor to provide a direct feed into the screening equipment.

Malcolm Construction has served the construction industry for more than 40 years by providing services in all areas of this sector, including civil engineering, earthmoving, plant hire, sports surfaces, waste management, recycling and quarrying, to help provide a 'one stop shop' approach for its customers. In addition, it is renowned throughout the UK as a specialist

A large and well laid out cab – as used on the 91-tonne TR100 mining truck.

A Komatsu WA500 is loading the screening equipment to sort the material into different sizes.

Quick tip cycle times from the twin inboard, two stage hoist cylinders and a heated body to cope with the freezing wintery conditions. Note the exhaust flange at the front of the skip.

One of the two Komatsu WA500 wheel loaders seen here on stockpiling duties.

A JCB JS360 excavator loads Terex mobile Pegson jaw and cone crushers from an elevated position, with the two crushers working in tandem to produce specified quality materials.

contractor for indoor and outdoor leisure pitches that can be constructed using Loanhill's quarry products.

As an update, I returned to this quarry during November 2013 to find the needs of the business had changed, by reverting from using a rigid quarry truck back to its tipper fleet. New crushing equipment had been installed at the mid-level benches and the two Komatsu wheel loaders had also been replaced – as part of Malcolm Construction's continuous investment in the quarry – with a Cat 980K and 972K wheel loaders under Finning's Just Add Diesel deal. In order to avoid duplication of information on Cat wheel loaders, these Cat products are covered in detail in Chapters 22 and 23.

Taken in November 2013, this 31-tonne Cat 980K is sporting a 5.7 cu.m bucket, which is capable of loading the 20-tonne capacity eight-wheeler trucks in just two passes.

CHAPTER 16

Lafarge Cement UK • Komatsu PC2000-8 excavator • Dunbar Quarry • July 2012

The background to Dunbar is interesting; since the first kiln was lit on 1 April 1963, the company has been supplying Blue Circle brand cements to customers in bags and in bulk and is the only cement works in Scotland. The factory and adjacent quarry is located on the east coast, 30 miles east of Edinburgh, between the main London–Edinburgh rail line and the A1, and has a good view of the North Sea.

The site has capacity to make more than 1 million tonnes of cement a year, principally for distribution to customers across Scotland and northern England. Lafarge claims that 6 tonnes of cement is used in building the average house in the UK and that its cement has also been used in large scale building projects which include: Carnoustie Golf Clubhouse and Hotel Complex, Royal Scottish Museum, Scottish Parliament building in Edinburgh and the impressive Falkirk Wheel Shiplift. The adjacent quarry that supplies the factory with the main raw materials of limestone and shale is the second largest quarry in Scotland.

Lafarge has made substantial investments in Dunbar Works in the last 25 years. A £40 million programme in the mid-1980s was followed by a £35 million upgrade in the mid- to late-1990s to increase production capacity and efficiency, improve safety and environmental performance.

Lafarge Cement UK has invested in a 200-tonne Komatsu PC2000-8 backhoe prime mover and four 91-tonne capacity Caterpillar 777F dump trucks at Dunbar Quarry. The investment at the quarry for front line equipment was expected to be around £5 million during 2012, which is part of a contract awarded to Babcock International Group. This represents a £100 million contract over the next ten years to provide a fleet-managed service for its Heavy Mobile Equipment

(HME) located across its aggregate and cement sites in North America and a £50 million fleet management contract awarded to Babcock in July 2011 that covers Lafarge's aggregate and cement sites in the UK.

Lafarge is looking for Babcock to deliver a whole lifecycle approach to managing its fleet of vehicles and equipment, from procuring and supporting the equipment through to arranging equipment disposal. The ten-year contract covers loaders, heavy rigid and articulated trucks, excavators and bulldozers; some 400 assets across 70 aggregate and cement sites in the UK.

Dunbar quarry has been operating some old equipment for some time and this new finance deal has allowed the site to procure a Komatsu PC2000-8 to replace an ageing Liebherr 994 backhoe, and four new Caterpillar 777F dump trucks, which will be covered in detail in chapter 17.

Lafarge's Komatsu PC2000-8 is the first to operate in Scotland and is the third machine to be operated in the UK. The other two PC2000 backhoes are earning their keep in a Welsh coal mine. The Japanese-designed machine was delivered and built on site by the UK dealer Marubeni–Komatsu from its Hamilton branch near Glasgow. It took eight articulated lorry loads and a team of four skilled fitters working ten hours per day over nine days to assemble the PC2000-8. Quarry manager Dave Hurcombe was really impressed with how the PC2000-8 was delivered in modules and the way the Marubeni–Komatsu team conducted itself on site. The team also came equipped with a comprehensive and robust method statement to build the machine safely.

Safety is the number one priority for the employees, contractors and visitors to any Lafarge Cement UK site; the working scheme ensures that areas that have been worked are restored promptly. The quarry uses an efficient opencast cut-and-fill technique to ensure the overburden is not double handled. Some 43 million tonnes of limestone are still to be extracted over the next 30 years. Current overburden ratios are 2.5:1, and with the stratum on a downward slope this will increase to 3:1.

Before the prime movers can get their teeth into the valuable limestone, more than 120 million tonnes of overburden will need to be shifted to unlock the two bands. One layer is 5m in depth and the second band is 7m thick, with 10m of sandstone between the two layers, so there is going to be some tough digging and loading in store for the big 200-tonne backhoe machine. The PC2000 is normally deployed to work a 5m high bench, whereas the Demag H135S face shovel is used on higher faces up to 11m in height. Going forward, the old Demag will be replaced by another face shovel to give the operation this flexibility – the replacement machine will be covered in detail in a later chapter.

Blasting is carried out on a regular basis and while Lafarge contracts out the drilling of blast holes, since it operates close to the main railway line it prefers to use its own team for blasting as tight control of blasting times has to be co-ordinated with Network Rail.

During my visit the enormous 200-tonne PC2000-8 backhoe had finished the overburden phase at the dig area and was working a 5m high bench loading high quality limestone into the 91 tonne capacity skips, and the machine will also be deployed for soil stripping work. UK-based MST manufactured the bespoke 12 cu.m bucket entirely of Hardox steel and has Hensley XS145 teeth fitted. With a 110% fill factor, the bucket will load the 777F in just four passes, creating a good heaped load, as experience has shown that attempting to add a fifth pass creates unwanted spillage around the site.

As Komatsu claims, and my first-hand experience confirms, for the size of the machine the exterior noise is remarkably quiet! This is attributed to the power module, which features large sound absorbing fan blades that help reduce environmental noise to an incredibly low level.

At the heart of the PC2000-8 is a powerful 30.4 litre, 12-cylinder twin-turbo with after-cooling, producing an impressive 956hp from the Komatsu SAA12V140E

A Demag H135S face shovel is used on higher faces up to 11m in height, seen here loading two Cat 777F trucks.

engine. This combines with Komatsu's Total Power Management system (which minimises power losses in the hydraulics, cooling fan and PTO) to dramatically reduce fuel consumption. Based on more than 700 hours of operation, Lafarge's PC2000 is consuming diesels at an average rate of just 90 litres per hour. That is 30 litres per hour less than the old Liebherr 994, and it is also impressive considering the 300-tonne class, monster miner, fuel burn will normally be about 170 litres per hour, resulting in a significant cost saving.

The PC2000-8 engine is environmentally friendly too, as it complies fully with Stage II emissions regulations. For additional fuel savings, the machine is fitted with an idling caution 'eco-gauge'. Lafarge has also set up the machine to automatically shut the engine down after three minutes of idling and the adjustable 'E mode' will help the operator work in the most fuel efficient manner possible, with the mode selection made via the in-cab 7in touch screen monitor.

Reducing operating costs even further, the PC2000-8 offers easy maintenance and repairs as its new design uses fewer components, helping reduce parts costs. The engine, radiator, oil cooler and PTO are all packaged in the 'power module', making it easier to remove and install components. A catwalk around the power module and centre walkway provides easy access for pre-use checks, inspection and maintenance. Furthermore, the Vehicle Health Monitoring System (VHMS) monitors all major components and enables remote analysis of the machine.

Komatsu claims that the cab design on the PC2000-8 is extremely safe and comfortable for the operator, is 30% larger than previous models (PC1800-6) and offers an excellent all-round view. Its sturdy construction was specially designed for mining applications (it is the same cab that is fitted to its German cousin, the Demag-inspired 300-tonne Komatsu PC3000-6 mining shovel) and includes a top guard that conforms to OPG level 2. Lafarge's machine is fitted with a rock guard across the large, full-height, front window, with the horizontal guard plates tilted up (on the top half) and downwards (on the bottom half) to align to the driver's eye line to ensure maximum visibility. Catwalk grilles around the cab have the same feature so the driver has good visibility through to the ground level.

Access to the cab is via the nearside, centre-mounted,

hydraulically activated drop down steps, again a similar design to that found on the PC3000-6 model. From an operator and safety perspective, it is this type of safety feature, along with all the working at height issues that have been addressed by well-designed catwalks and handrail, which makes all the difference when it comes to selecting a machine. Dave Hurcombe explains: "Lafarge take health and safety very seriously and are looking at continuous improvement wherever possible, so we welcome such a well-designed machine, packing lots of safety features into our prime mover fleet."

Sitting comfortably in the air suspension seat, the dual rear-view mirrors help reduce blind spots, while Lafarge specified the optional rear-view three camera monitoring system, fitted to the n/s/r, centre rear and o/s/r to cover all possible blind spots via a 7in TFT-LCD colour monitor. One slight disadvantage of this set-up is the screen is also used, in the main, to monitor the machine's performance and its touch sensitive screen is also used to change engine power modes. That said, the screen is quickly switched to a split screen displaying two camera views when the PC2000 is carrying out any manoeuvring.

From the trainer seat, I had the rare opportunity to experience the PC2000-8 being put through its paces, and if the low exterior noise was good, then the interior noise levels are incredibly low, at a claimed rating of only 64.5dB(A). This is achieved with the help of six new cab damper mountings, in combination with power module design. That is the same noise levels of a modern passenger car travelling at 50mph!

Having spent thousands of hours on the old Liebherr 994 backhoe, PC2000 driver Adam Reid is really impressed with the new machine, and comments: "It is a great digging tool, and it's a good place to spend an eight-hour shift, I like everything about it, from the very quiet, comfortable and air conditioned cab, to the smooth and powerful hydraulics. And, having driven excavators for more than 30 years, this new Komatsu is in a different class. It loads the Cat trucks in four quick passes, is very stable and daily checks are easily carried out with wide catwalks around the engine compartment."

Adam has only two minor reservations about the machine, the first being that it has a slightly shorter reach compared to the Liebherr 994. The second is that the

With a bucket fill factor of approximately 110%, the PC2000 loads a 777F in four quick passes.

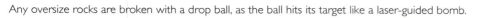

Any oversize rocks are broken with a drop ball, as the ball hits its target like a laser-guided bomb.

Sitting on top of a 4m high blasted rock pile, the standard boom and stick on the PC2000 has the required breakout forces to load efficiently.

Hydraulic-operated access steps to safely reach the cab and upper structure.

main work-lamps are fitted to the front, under the cab, but given the ease of access and cavernous space inside the cab base structure, Adam thinks the lamps should have been hinged to aid cleaning. He has a point and that is one for the Japanese design team to consider during any upgrades.

When it came to specifying the machine, Dave ordered the PC2000 with a standard boom and stick; to have ordered the machine with more reach would have adversely affected the breakout forces and would have further compromised productivity by fitting a smaller bucket. And while the quarry team blast and obtain good rock fragmentation, they needed a machine that could dig hard and the PC2000 has more than matched their expectations.

Dunbar quarry also operates another Komatsu, commissioned in 1999, the Demag-designed H135S face shovel. This machine had an extensive overhaul carried out costing nearly £160,000 that covered fresh hydraulics, re-cutting the main bushes and the fitting of a

Good visibility into the back of the truck from an elevated position, along with side and rear cameras fitted and linked to the in-cab LCD monitor.

PC2000 seen here soil stripping at the top of the quarry. Any soil extracted will be used for reinstatement later on.

The 12 cu.m bucket is so large that a massive piece of oversize stone has inadvertently been loaded into the 777F skip. This will be taken to the Demag H135S to be broken down with a drop ball before transporting it to the primary crusher.

new Cummins engine. The 'old girl' looks in reasonable condition, apart from a number of track pads missing, as some parts are getting difficult to obtain now. Dave was going to try some locally made pads fabricated from Hardox steel as a stop gap until a replacement machine was delivered during 2013.

Since the quarry's primary crusher can only handle limestone rocks that are no bigger than a square metre in size, the skilled driver of the H135S uses a drop ball to break the larger pieces of stone with the effectiveness and accuracy of a laser-guided missile, in this case without the 'shrapnel'!

With the PC2000-8 backhoe now firmly established with operators and management, Dave had a twinkle in his eye for a PC2000-8 face shovel to replace the old H135S face shovel. This proved to be the case, as I returned to see it being built and working in 2013. This is covered in detail in Chapters 19 and 20.

A bespoke MST-designed 12 cu.m rock bucket with heavy duty wear package fitted.

CHAPTER 17

Lafarge Cement UK • Caterpillar 777F haul trucks • Dunbar Quarry • July 2012

Following on from the previous chapter looking at the PC2000 at Dunbar quarry, I also looked at the Caterpillar 777F haul trucks there at the same time. As part of a £5 million investment at the quarry, four new Caterpillar 777F haul trucks are a welcome sight to both drivers and management alike, as they have replaced four five-year-old Terex TR100 trucks that have come to the end of their lease agreement.

Dave Hurcombe explained that the TR100s needed to be replaced, and having reviewed the market for rigid 91-tonne trucks, the company decided on Caterpillar's legendary 777 haul truck as its best overall option, and was happy to select the F model, due to shorter lead times, rather than wait for the newer G model.

Dave is delighted with their performance and said: "I had three problems to deal with when the first batch of two Cat 777F trucks arrived on site; one, the drivers all wanted to drive them, secondly, the reception on the radio and CD player needed fixing, and last but not least, the cabs on the trucks are so well insulated from exterior noise that the drivers were having difficulty in hearing the excavator horns to signal their final pass is in the skip.

The trucks arrive at the primary crusher and wait until they see the green light before tipping their 91-tonne load of limestone.

"Both Finning and I were really pleased that's all we had to deal with. Finning support is great and they quickly dealt with the radios. The drivers are happy now that I've got my other two trucks on site, and the cab being so quiet, well, that's a nice problem to have, so I'll just need to find better alert system for the excavators."

As mentioned in the mining chapter, with the Bell B50D trucks, original equipment manufacturers should look to address this issue by fitting louder horns or consider the system Banks Mining has devised at its Rusha site by fitting lights to its excavators that signal when a truck driver is good to go!

Lafarge and Dunbar are no strangers to Cat equipment as they still operate a tidy example of a big Cat 992C model wheeled loader and a 13-year-old D10R dozer. And Dave values Finning's service support and personnel, such as the fact that the same field service engineer has been looking after its Caterpillar equipment for the last 15 years. Dave feels it makes such a difference to have longstanding business relationships with the same people, because when they call Finning it knows the site so well that Dave does not need to explain the impact on production if he has a particular machine issue.

Prime mover availability is key at Dunbar quarry as the site only has a four-day buffer of high quality limestone stockpiled. And since the adjacent cement factory works through the weekend and the quarry does not, this means the cushion has been reduced to just two days' stock when Dave and his quarry team arrive at 6am Monday morning!

The 777F trucks will be playing their part over the next 28,000 hours to haul valuable limestone and help move some 120 million tonnes of overburden to unlock the two 5m and 7m deep bands of stone. And, as covered in the previous chapter, the four 777F trucks are matched to a new Komatsu PC2000 backhoe and old Demag-inspired Komatsu H135S face shovel, the PC2000 is swinging a 12 cu.m bucket and the H135S has a 9.5 cu.m bucket and can four and six pass load the trucks respectively. This creates a good heaped load before the 777Fs make the short 500m journey to the primary crusher, where they wait before being allowed to tip into the hopper.

Finning has fitted a Hardox steel liner package to the 60 cu.m/91-tonne capacity skips, to help protect the bodies from excessive wear and tear. The 777F skips

When the quarry needs some extra production, or when a prime mover is down for maintenance, it still operates a classic Cat 992C wheel loader from the days when the site was owned by Blue Circle cement.

provides a big target area for the two prime movers, aided by a red arrow marked on the side of the skip to show the excavator operators the centre point to load. This helps with fast hoist times and, when loaded correctly, puts less strain on the hydraulic hoist cylinders.

When the 777F trucks are hauling limestone, the trucks are not being used to their maximum efficiency as both excavators are normally teamed with only two trucks each because they have to blend two types of limestone together. However, the main production priority is keeping the 1,250-ton per hour primary crusher hopper full for maximum output. Dave explained that the new 777F trucks have on-board weighing and that he can also download this data from Finning – useful as one of his main key performance indicators is the output of the primary crusher. He is delighted to report an increase of 150 tonnes per hour from the Cat 777F trucks. This increase is down to a number of factors; the drivers like the trucks and are more productive as a result, and the other main reason is the 777F has good gearing and a traction control system that is well suited to this application.

At the time of my visit the main haul road was in good condition as it was covered with a layer of extracted sandstone. However, the same could not be said for parts of the dig area due to an exceptional amount of rainfall and because the site has limited capacity to deal with maintaining both the haul road and dig areas with a single Cat D10R dozer available. That said, the 777F trucks were coping with the challenging ground conditions, which was just as well, as the trucks needed to muck-shift as quickly as possible to get to the limestone and keep the cement factory fed.

Gerry Donnellan is one of the 777F drivers. With more than six years' experience at the Dunbar quarry, he found the 777F trucks more comfortable to drive, with lots of space in the cab, particularly for some of the taller staff. It also has a good gearbox producing really smooth shifts, holds the gears well and has lots of power and traction to climb the steep ramps out of the quarry floor.

The cab interior is spacious, is fitted with a comfortable air suspension driver's seat, and has good visibility out of the large front screen. All-round visibility is enhanced with the fitment of sophisticated front and rear cameras and sensors, with the colour monitor positioned up in the right-hand side of the cab. Also adding to the all-round visibility are the large and well-placed mirrors fitted to the front as well as the heated mirrors on each side of the truck, six in all. Caterpillar claims the cab interior noise level to be just 76dB(A); not bad for a truck producing more than 1,000hp and hauling 91 tonnes of material.

With health and safety at the top of Lafarge's agenda, the 777F trucks are spec'ed with some interesting optional extras. These include Caterpillar's object detection system, whereby the truck is fitted with radar sensors front and rear that are designed to enhance the driver's awareness during the critical period when he is starting the truck and during the first several seconds of movement. The system consists of an interactive touch screen display, short range radar (up to 7m), medium range radar (up to 20m) and cameras on the front, rear and sides of the machine to help prevent these blind spots, creating a safer environment for site personnel and equipment.

Caterpillar's 777 models are a popular truck in the coal mining sector, with most operations choosing to build berms so the drivers can access the front bumper catwalk without the need to use the flexi-steps that hang below the bumper. Lafarge has taken a different approach and chosen to fit a Cat-designed, hydraulically operated, drop down and angled step arrangement, making access to the cab safe and easy for the drivers to use.

While the site enjoys the longstanding support of the local Finning engineer, Lafarge takes full advantage of Finning's Finslight service – a satellite-based Vision Link system, whereby highly experienced technical staff back at Cannock, Finning's HQ, can monitor every performance sensor on the new trucks and interpret the information displayed, just like an F1 team. So, if the need arises, the staff can alert the local field-based engineer to check out any suspect parts that have been flagged up. The 777F trucks have increased service intervals to every 500 hours (250 hours on the previous 777D model) along with oil sampling to help ensure early warning of parts that need replacing before failure.

At the heart of the 777F beats a powerful 32.1-litre V12 cylinder, twin-turbo with after-cooling, producing an impressive 1,016hp and 4,716Nm of torque at 1750rpm from the CAT C32 ACERT engine. Based on more than

The Cat 777F making its way from the soil stripping area. Note the Dunbar lighthouse in the background looking out to the North Sea.

The Cat 777F trucks have an upgraded traction control system to deal with adverse ground conditions and save on tyre wear.

1,200 hours of operation, Lafarge's Cat 777F trucks are consuming diesel at an average rate of just 54 litres per hour – that's 5 litres per hour fewer than the TR100s – and if fuel burn remains constant over the operating lifetime of the four trucks, this will save the quarry some £400,000 on its fuel bill!

The four 777Fs are protected from potential fire damage as they are fitted with an optional extra fire suppression system, covering the engine and drive train areas. Wearing components are also protected by the autolube system, supplying grease to the steering linkage, suspension units, hoist cylinders and skip pivot points.

Like the PC2000 excavator, the four trucks were procured through Babcock International group, and the deal will also cover planned overhauls on the 777F trucks at around 17,000–20,000 hours depending on the preventative service information. The only costs that are not fixed are wearing parts, such as tyres that cost in the region of £55,000 to replace all six! And once the trucks reach 28,000 hours, it will be time for Lafarge and Dave Hurcombe to carry out a market review again and select replacement dump trucks.

Optional extra Cat-designed drop down access steps are fitted to provide much safer access to the cab for the driver.

CHAPTER 18

Hillhouse Group • Cat 390D LME and 775G • Hillhouse Quarry • November 2012

Hillhouse Quarry, part of the Hillhouse Group, has remained a family-owned company, with a long history of producing crushed rock aggregates, since 1907. When it started out, the plant was capable of producing 150 tons of stone per shift (about 35,000 tons per annum) using the techniques of the day, which included a rail track and waggon system from the face. With the advent of modern machines and technology, it now has the capacity to deliver around 1 million tonnes of drystone per annum and other related products for the construction industry. The quarry has planning permission for activities to continue until 2042 and is located on the A759 to the immediate south-west of Dundonald, near Troon, South Ayrshire.

Hillhouse Quarry management recently carried out a review of its frontline excavator and haul truck team, with the objective to extract high quality whinstone at the lowest cost per ton. A number of manufacturers and

The Cat 390D loading a 775G, with the rest of the quarry operation and the beautiful island of Arran seen in the background.

dealers submitted proposals to cover machine whole life cost, over a five-year term or 10,000 hours of operation, with front end price and warranty, the other two key performance indicators being residual value and fuel burn.

The proposals on the table were whittled down to just two, with Caterpillar's UK dealer, Finning UK Ltd, producing the most compelling deal, backed up by manufacturer's stated performance figures and historical evidence of Caterpillar equipment enjoying strong residual values. Finning was also able to produce anonymised real-life performance data for a 90-tonne class Cat 390D LME excavator and a 70 tonnes class payload Cat 775G haul truck to convince management its equipment would meet all the objectives.

The other main factor in the buying decision was aftercare service and that is where Finning scored highly too, as Hillhouse quarry has been operating Cat equipment for a number of years and has enjoyed good levels of service during this time. Response times have typically been around one hour from making the call to the Cat engineer being on site. Justin Gill, operations manager at Hillhouse quarry, explains: "Finning service is outstanding, as our longstanding Finning engineer Calum White really takes ownership when the Cat equipment needs attention and he went the extra mile for us, which Finning recognised by awarding Calum with employee of the month."

Hillhouse Quarry management finally decided to invest £1.5 million in two Caterpillar 70-tonne capacity quarry trucks and Caterpillar's top of the construction range 90-ton excavator, which are expected to deliver efficiencies at the quarry face and across the business.

Prior to the arrival of the 390D excavator and two 775G trucks, the quarry was operating a 66-tonne, 404hp Cat 365C LME and ageing Bell 30- and 40-ton capacity ADTs to load and haul the blasted stone to the primary crusher. In order to achieve the lowest cost per ton objective, the two 775G trucks have allowed the quarry to reduce its truck fleet from seven to five.

Before the hand-picked team of Andy Galloway operating truck number two, John Dickie operating the 390D and Ian Barbour driving truck number one were let loose with their new tools, Hillhouse quarry management asked Finning to carry out refresher training thorough its eco drive course. Gary Jones,

quarry manager, explained; "Although the guys are very experienced operators, having spent many hours driving ADTs, we felt that the team would benefit from Caterpillar's eco driving course using rigid trucks and they were surprised how much they learned about operating the trucks and excavator, and picked up useful tips to improve fuel consumption."

Andy Galloway, the proud driver of truck number two, has previous experience of driving 91-tonne class trucks from both Caterpillar and Terex. He comments: "The 775G trucks are really nice to drive, with spacious, quiet and comfortable cabs and smooth transmission. The engine is powerful and quickly gets up to fifth gear, and due to the short distances to the primary crusher we don't need to use the remaining top two gears."

John Dickie, operating the 390D LME (in mass excavator configuration), can load the 775G with 70 tonnes of rock in six passes and under one minute forty-five seconds. He comments: "The 390D is a big heavy machine as it's very stable and planted, and the large counterweight helps to ensure there is no nodding when digging hard. The 365C was a good tool, but this 390D is something else – very powerful and absolutely effortless in loading these 70-tonne rigid trucks."

Having had previous experience with older models of Cat's 775 quarry truck, fitted with Michelin tyres, Gary Jones specified Michelin's X-Quarry S tyre and is expecting to get about 6,000 hours from them. Michelin claims these tyres have the deepest tread in the market, lasting 15% longer than the Michelin XKD1, and they have a reinforced protective band designed to be resistant to cuts and scrapes in abrasive quarry operations.

Caterpillar claims the 775G series truck is backed by comprehensive validation through its 'field follow' programme of prototype testing with some 33,000 hours of work in the hands of customers prior to production. It has a lot of additional safety features fitted as standard, including new service braking performance, with engine braking options for automatic retarding control. Also, the driver's seat is a series III model with vibration reduction that adjusts to individual suspension needs and has a three-point seatbelt.

The cab has automatic temperature control, electric window (left side), and emergency egress out the right side, hinged, window. It has improved ground level daily

One of the Cat 775G trucks heading to the crusher.

Excellent view out the large front windscreen.

Spacious cab with trainer seat.

checks and a redesigned tread plate on the cab access steps for better underfoot grip.

The truck also has a sophisticated tyre monitoring system, whereby it takes the payload and distance value and combines it with ambient air temperature, machine speed and the manufacturer's rating for the 775G's tyres to calculate tyre condition continuously. As a tyre approaches its temperature limits, the operator gets a warning inside the cab to prevent reverse vulcanisation of the tyre and associated damage. Caterpillar claims this is an exclusive feature, and it is an important tool to help extend tyre life.

Quarries around the UK take health and safety very seriously and none more so than Hillhouse. Gary is pleased that Caterpillar has redesigned the 775G model to be serviced at ground level as this minimises working at height issues with the engine's main filters easily accessible.

Other safety equipment specified on the 775G includes a radar warning system by Ogden Safety Systems Ltd. The FMCW radar is designed to address two conflicting requirements: it is designed to respond to objects that present a hazard but not 'cry wolf' about objects that do not. The system combines with Caterpillar's excellent rear-view camera and LCD monitor to give the best of both worlds, with an audible and visual warning to the driver when a real hazard approaches.

Orange strobe lights are fitted in the truck's radiator grille. Other additional equipment consists of spot lamps – fitted close to the heated mirrors – to illuminate the n/s/f wheel area at night. Flashing green warning lamps, to indicate when the driver is using his seatbelt, are fitted to the trucks and the 390D excavator. Gary also reports the main headlamps on the 775G are much improved on the previous model.

Safety equipment on the Cat 390D consists of a rear-view camera fitted as standard and the LCD monitor has a multi-function display. In addition, the large rear counterweight has been painted with highly visible red

and white chevrons. With the quarry face at the dig area a place where there is a foreseeable risk of falling rock or collapse, Hillhouse quarry has fitted its Cat 390D with Cat's FOGS (Falling Object Guard System) including overhead and windshield guards. The cab structure allows the FOGS to be bolted directly to the ROPS cab. Management has also specified full-length handrails to the standard 500mm wide catwalks on both sides of the upper structure to allow safer access to the fuel tank cap and access doors to carry out maintenance tasks.

The Cat 390D cab shell is attached to the frame with viscous rubber cab mounts, which dampen vibrations and sound levels, and has electronic control joysticks – on the air suspension seat – that Caterpillar claims reduce noise levels over hydraulic pilot lines. Collectively all these features enhance operator comfort and help to reduce interior cab noise levels to just 74dB(A).

At the heart of the 390D is Caterpillar's C18 ACERT engine, fitted with a uniquely designed water-cooled turbocharger and mechanically actuated fuel injection that produces 523hp from its Stage IIIA compliant six-cylinder 18.1-litre capacity. The 390D L optimises fuel consumption through flexible power settings, high production or economy mode – set on the LCD screen – which electronically manages engine response to match load and demand. The 390D has SmartBoom technology and works well when loading trucks from a bench position. This feature is claimed to make the 390D more productive and fuel efficient as the return cycle is reduced, while the boom down function does not require pump flow.

The 390D L in ME configuration is fitted with a 7.25m mass boom and 2.92m short stick option and is swinging Caterpillar's 5.7 cu.m (weighing 8 tonnes) extreme duty rock bucket. This has been designed for highly abrasive conditions, with corner shrouds added, along with larger side wear plates for added protection.

Hillhouse quarry has a blast-to-load strategy that obtains high fragmentation to achieve minimum oversize rocks in order to keep the primary crusher working efficiently. During my visit the 390D was sitting on a 3.5m high bench of newly blasted rock and firmly planted with its 650mm wide double grouser shoes. It looked effortless as it quickly delivered six passes into the back of the 775G 42.2 cu.m, 71-tonne capacity quarry spec-body. When fully loaded, the load warning

system light, fitted to the n/s/f, turned to red. Hillhouse has asked Finning to preset the load warning system to 70 tonnes in order to produce a good heaped load, while keeping overspill to a minimum.

Once the driver gets the toot from the excavator horn, to signal the final pass has landed in the 775G skip, the big quarry truck pulls easily away using its Cat C27 ACERT diesel engine, which is capable of delivering an additional 5% more power to the drive train than the previous 775F model. As a result, the transmission has more robust components; including a larger driveshaft and differential gears to handle the 825hp at 2000rpm from the 27-litre V12 twin-turbo engine.

The 775G series truck has a new planetary powershift transmission control strategy that takes the benefits of an electronic clutch pressure control system and adds part throttle shifting and torque shift management. The result is smooth automotive-type shifting and retains torque and momentum through the shifts; increasing performance on grades. The 775G also has a new traction control system that is now steering sensitive to differentiate between tyre spin and high speed turns.

Much like the Cat 390D, the 775G cab is claimed to have an interior noise level of just 76dB(A), a claimed 50% reduction from the previous F model, and the same performance as its bigger brother, the 777F. At a quick glance you could be forgiven for thinking the 775G is as large as a 91-tonne class truck, as the 775G is only 500mm shorter in overall length than a 777F, at 10.07m! The view from 775G cab provides excellent all-round visibility and has a good, heated, mirror package to the front as well as on each side of the truck, four in all.

Once loaded the 775Gs make a short 494m trip to the 600 tonnes/hour Allis Chalmers 42/65 primary crusher and can quickly hoist its skip to discharge its load in around 9½ seconds.

Once the rock is crushed, screened and stockpiled, Hillhouse quarry Cat 980H wheeled loaders are used to dispatch material into one of the contract hauler's HGV eight-wheelers. The 980H loaders are also required to travel up to the spotting area every two hours to tidy up; this procedure helps to protect the truck tyres from sharp rocks when they reverse back to be loaded.

A £2.5 million investment in new secondary crushing and screening plants during 2009 has boosted total production to around 1 million tonnes of drystone per

One Cat 775G returns to the spotting areas while the other waits for it to pass safely.

annum. With sufficient reserves of stone to last well in excess of 20 years at current levels of production, this recent upgrade at Hillhouse Quarry represents an ongoing and proactive programme to modernise and improve the entire operation of both fixed and mobile plant to ensure the site continues to meet future demands in the most sustainable, efficient and effective way possible.

Hillhouse management has access to real time data via Caterpillar's on-board Product link system and web-based application called VisionLink. Using this system, based on 100 hours of operation, the Cat 390D is burning fuel at a rate of just 34 litres per hour and the Cat 775G is consuming diesel at an average rate of just 30 litres per hour. That is a significant litres per tonne advantage from operating two ADT with the same payload capacity as just one 775G, and if fuel burn remains constant over the operating lifetime of the two trucks, this will save the quarry significant costs on its fuel bill.

Hillhouse quarry has fitted its Cat 390D with Cat's FOGS (Falling Object Guard System), including overhead and windshield guards.

The 775G is nearly as large as a 91-tonne class truck, as the 775G is only 500mm shorter in overall length than a 777F, at 10.07m!

Cat 390D excavator was sitting on a 3.5m high bench of newly blasted rock with its 650mm wide double grouser shoes, and looked effortless as it quickly delivered six passes into the back of the 42.2 cu.m skip.

The 775G truck makes a short 494m trip to the 600 tonnes/hour Allis Chalmers 42/65 primary crusher and can quickly hoist its skip to discharge the load in around 9½ seconds.

Cat 980H loading type 1 stone into a contract haulier's eight-wheel tipper truck.

CHAPTER 19

Lafarge Tarmac • PC2000-8 face shovel build • Dunbar, July 2013

As covered in Chapter 16 on Dunbar cement quarry, Lafarge Tarmac is operating an old Demag-inspired Komatsu H135S face shovel that is due for replacement. And I make a return visit to Dunbar to see the first Komatsu PC2000-8 face shovel in Europe being built by UK dealer Marubeni–Komatsu Ltd (MKL) to replace this ageing shovel.

The PC2000-8 face shovel (FS) started its journey from the Osaka plant, Japan, where the components are manufactured. It travelled more than 13,000 nautical miles on a two-month journey to the Port of Antwerp, Belgium, where the PC2000 parts made their onward final sea journey to Newcastle. WWL ALS (a division of Wallenius Wilhelmsen Logistics) co-ordinated the delivery, as they had also moved the three previous PC2000s operating in the UK. WWL ALS team preparations included completing a detailed and comprehensive load plan and deployed ten trucks for the move. The largest part of the machine was the main frame of the excavator weighing 31 tonnes and measuring 3.4m wide. The move was supported by Marubeni–Komatsu's own HGV drawbar trailer combination to pick up some of the main components and bring them back to Dunbar, with the final mile covered using a section of the old A1 road, now only used as an access point to the southern end of the quarry.

Dave Hurcombe, Lafarge Tarmac quarry manager, had the site prepared when the MKL team built the quarry's first PC2000-8, in backhoe guise, and made sure the site was still in good condition and suitable for this build.

Kevin Henderson, Marubeni's lead field service technician, was the person responsible on site for the safety, quality and performance of the four-man team

that would be building the PC2000-8 FS. Kevin also built Dunbar's backhoe machine. Brian Watson, field service technician, along with his stunning £200,000 support truck, has helped build all three PC2000 excavators operating in the UK, and this one takes his tally to four. Other members of the team include Craig Eldridge, David Balfour and second year apprentice Paris Stone.

The build starts on Monday, 8 July with Kevin and his team setting up a self-contained office–welfare container unit and separate parts store, while they await the convoy of trucks from Newcastle. They have also hired in a telehandler, cherry picker and a calibrated airgun torque wrench, which are checked over and any slight damage, to tyres, etc, is photographed and logged.

The area is coned off around the build area; including a side entrance where the old Cat 992C wheel shovel is loading stockpiled material into Cat 777F trucks. All the other health and safety paperwork, including risk assessments, are on display for Dave Hurcombe to examine when he visits to check on progress. While MKL has comprehensive risk assessments (RA), Kevin and his team spend half an hour at the start of each shift writing out a specific RA for that day's activities and this also allows them to safely focus on a series of sprints each day, over the next 14 days of the build.

Tuesday, 9 July. With the HGV trucks parked up overnight on the access road, the two 100-tonne capacity Terex–Demag mobile cranes duly roll in at 8am to start unloading the PC2000 parts and carefully position each of the components around the build area as per Komatsu's assembly manual for this 200-tonne monster miner, as part of the pre-planning carried out by Kevin.

Brian's support truck, with its 50-tonne rated capacity vehicle-mounted crane, helps to unload and makes short work of moving the PC2000 9,000kg dipper arm. By 10am the left-hand track frame is off the low-loader trailer, followed by the centre section car body of the undercarriage. These two components are carefully bolted together using 13 bolts from (inside bottom and side) and a further six bolts (top outer side) of the track frame. The team uses its calibrated airgun – fitted with a torque multiplier – and begins to tighten all 38 × 45 mm diameter bolts to an impressive 4,900Nm of torque. In the afternoon the right-hand track frame swings into place and a further 38 bolts are fitted and torqued up. The most important job after this point is to make sure the undercarriage is positioned exactly dead level by using wooden blocks under the track shoes; otherwise the team will struggle to get the large mainframe fitted easily the next day.

Wednesday, 10 July. At 7am the day starts with the team preparing the upper structure main frame (rotating frame in Komatsu speak) by carrying out a number of jobs while it is still on the ground (to minimise working at height issues) such as grinding off any paint from areas that are going to be bolted together. If this is not done the paint can be compressed and components are then prone to loosen off, such as the cab base unit and side catwalk panel mounting points.

The swing-ring gear is filled with grease and at 9.30am the two Demag cranes start to lift the mainframe and position the two swing motors to mesh with the swing gears and swing-ring dowel pins. At this point the mainframe is lowered carefully, and the team can start to fit the 60 × 36 mm diameter bolts that connect the two components, and are torqued to 3,040Nm. Once fully bolted down, the team can off-hire one of the two cranes. With good access, at this point, they finish off connecting pipes to the swing motors and centre rotary valve that supplies the travel motors with hydraulic oil.

The next big lift of the day is fitting the 2,600kg cab base unit on to the mainframe. By 4pm the left floor assembly, which includes the nearside drop down access steps, side panel and cab step assembly, have been fitted. Kevin and his team are working hard in temperatures nearing 30°C, but there are no complaints, as it's nice and dry with plenty of daylight to work with. Kevin has planned to get the 16,200kg power and hydraulic pack on before they finish for the day. Komatsu has designed the mounting brackets on

the mainframe with Vee lugs, so the power pack is easily guided into its exact position. The securing bolts are fitted just before finishing for the day. Meanwhile, Brian and Paris prep and lift the o/s/r side panel and catwalk assembly, with Kevin and Craig installing the bolts as it is swung into place. Working 12-hour shifts has paid off, as the PC2000-8 is really starting to take shape in just two days of hard graft.

Thursday, 11 July. With good weather still holding, the team makes another early start by fitting the main hydraulic suction lines from the hydraulic tank to the pumps housed in the newly installed power pack. Next up is the n/s/r side panel and catwalk assembly. By 10am the massive 26,500kg counterbalance is lifted into position and Davy starts to push through the 12 large 42mm diameter × 670mm long bolts, six on each side of the mainframe leg beams, and using the air torque wrench, tightens up the bolts to 4,200Nm.

Meanwhile, Craig and Kevin are fitting the cab rear floor assembly, which sits on top of the hydraulic tank, and the cab outer catwalk and handrails in preparation to lift the cab. With this in place, it provides a safe working area and platform to fit the large cab, from the Komatsu Mining Germany (KMG) PC3000-6 model range, and by mid-afternoon the 1,900kg cab is lowered on to the lower cab base unit and bolted down against the numerous rubber damper mountings. The rest of the day is spent fitting the exhaust stacks and air filter units, and breaking the boom, dipper and lift rams out of their wooden packing crates ready for tomorrow's next big lift.

Friday, 12 July. Throughout the build, Kevin's team has been wearing full working-at-height harnesses and clipped on to structural parts of the machine – just in case someone loses their footing – and this morning is no different as Craig is clipped on to the valve block as he sets about partially withdrawing the huge pivot pins (weighing 90kg!) from the boom foot tower. With the pins out of the way, the Demag crane lifts one of the 2,400kg main lift cylinders into position. Kevin comments: "These pins and bushes are so well engineered that with just a bit of copper grease you can just about push them home with hand pressure only." By 10am both cylinders are fitted and pins secured. The crane then swings the 11,000kg boom in front of the PC2000 so Brian Watson's Palfinger crane can lift the two smaller 1,072kg

stick cylinders and attach them to the boom with their mounting pins and temporary slings.

If you have read the chapters on mining, you may have noticed that the design of the PC2000 FS boom, stick and bucket is remarkably similar to that of its German cousin – the 300-tonne class PC3000-6. That is because, as I understand it, the front end equipment was designed by KMG and given its long and enviable reputation in the mining sector, this should prove to be a smart move by Komatsu Japan, as it has chosen – through cross-team working – to draw on decades of KMG and Demag design experience in large mining face shovel technology.

Kevin and his team are pleased to see that the extensive hydraulic pipework on the back of the boom is pre-installed at the factory. With both stick cylinders fitted, the crane can now lift the 13,144kg boom assembly into position on top of the boomfoot tower and by lunchtime the lift cylinders are also connected to the boom. Brian can now sling and lift the two bucket cylinders on to the top of the boom, and by 4pm the stick is lifted into place by the Demag crane. By 4.30pm, with the last big piece fitted, Kevin can now off-hire the other 100-tonne capacity crane. On Friday evening the team can look back at a near fully formed PC2000 FS; it has taken just four days to get to this point.

By 10am the left-hand track frame is off the low-loader trailer, followed by the centre section car body of the undercarriage. Photo: Kevin Henderson

In the afternoon the right-hand track frame swings into place and a further 38 bolts are fitted and torqued up. Photo: Kevin Henderson

The swing-ring gear is filled with grease and the two Demag cranes start to lift the mainframe and position the two swing motors to mesh with the swing gears. Photo: Kevin Henderson

Komatsu has designed the mounting bracket on the mainframe with Vee lugs, so the power pack is easily guided into its exact position.

Some of the MKL team, from left to right: Craig, Kevin, Paris and Brian.

On day three, the massive 26,500kg counterbalance is lifted into position and secured with 12 large 42mm diameter × 670mm long bolts. Photo: Kevin Henderson

On day four the crane swings the 11,000kg boom into position, ready to attach to the PC2000. Photo: Kevin Henderson

Brian uses his crane to fit the bucket cylinders on to the boom … Photo: Kevin Henderson

… and by 4pm the stick is lifted into place by the Demag crane. By 4.30pm, with the last big piece fitted, Kevin can now off-hire the other 100-tonne capacity crane. Photo: Kevin Henderson

Day five, Saturday, 13 July. Going forward, the team can start to focus on the hundreds of small jobs that need to be completed over the next few days.

Day six, Sunday, 14 July. Rest day.

Monday, 15 July–Thursday, 18 July. The team carries out numerous jobs including connecting up the autolube grease lines to the swing ring, and attaching the hydraulic pipes to the cylinders inside the track frames – that pressurises the track idlers to 200 bar (every time the right-hand travel motor pedal is depressed during normal operation) – to name but a few.

Some of the biggest jobs to be completed over these next few days are fitting all the hydraulic lines, from the small pilot control lines routed from the cab and cab-base, to the large lift cylinder hoses. Craig comments: "The machine is delivered from the factory with many

components having been tested for operation, pressures set and checked for any leaks on valves and fixed pipework; Komatsu fit these handy blanking plugs with test points. We can use these points to connect a drain hose to relieve the static pressure and drain off the remaining test oil without getting covered in oil or making a mess of the machine or the customer's site."

With so many electrical jobs to be finished, Kevin brings in David Watson, MKL electrician, to connect everything including the computer monitoring system set-up using a laptop; work lamps; three rear-view cameras; and quarry spec green warning lamp mounted on top of the cab.

To fill the machine with high quantities of fluids, the team uses the telehandler to lift a 1,000-litre IBC container and gravity fill the hydraulic tank. Kevin has

Three ultrafine hydraulic filters are designed to initially capture any microscopic debris. These will be changed to standard filters once it is operational.

The team grinds off any paint from areas that are going to be bolted together, otherwise it can be compressed and components are then prone to loosen off. The side panel is being fitted with the use of Brian's vehicle crane.

Factory-fitted test points and blanking plugs; great for draining factory test oil prior to fitting the hydraulic hoses without making a mess of the site or machine.

End of week one – the team can look back at a near fully formed PC2000-8 FS.

arranged with Dave Hurcombe to fuel the machine's 3,400-litre tank via the site's fuel bowser and the fast fill point. Before the engine can be run up to speed, Kevin and the rest of the team must first bleed air out of – under gravity pressure only – the three main hydraulic pumps, then crank the engine over for only 20 seconds, and bleed the system again and carry out a similar process at the pilot control valves, swing motors, etc, otherwise serious damage will be caused to these parts.

This procedure is all covered in the PC2000-8 assembly manual, however, Komatsu also cable ties laminated warning and instructions sheets to each of the hydraulic components that must be bled, and once completed the tags are removed. In addition to this, during the first few hours when the machines components are being tested, Kevin has fitted three ultrafine hydraulic filters designed to capture any microscopic debris; the filters are so fine they cannot be used when the machine is in full operation, as they would restrict its digging performance and must be removed and standard service filters installed before handing over the machine to Dave Hurcombe.

Friday, 19 July. With the engine fully up and running and the hydraulic systems checked for leaks and operation, track idlers are pressurised and the team are able to track the PC2000 forward to connect the nearby 11.5 cu.m capacity bucket. With the machine moved off its perfectly level position the guys deploy a bottle jack to steadily drive home the huge pivot pins that connect the bucket and stick. Once again, Brian's support truck crane was used by positioning the big bucket rams to align with the bucket pivot pins.

MKL is getting a good return on its £200,000 investment, as Brian skilfully manoeuvres part after part to the nearest millimetre using his remote hand control set, thus saving on crane hire costs, and last, but not least, where would you get such a skilled crane operator that knows how build a PC2000 and is able to work in such close quarters without damaging an excavator worth around £2.5 million!

Throughout our numerous visits to site it was good to witness a new technician being trained. Brian, Craig, David and Kevin – with a combined 74 years' experience – all took the time, despite their busy schedule, to explain to Paris what he needed to do next, why and, most importantly with safety in mind, he understood the risks involved before turning a spanner. This was in evidence when removing hydraulic blanking plugs from the pre-installed pipes on the stick. The four bolts holding two pipe flange clamps had to be removed in a certain order, so the flow of the oil was released in the direction away from Paris.

MKL subcontracted only two areas of the build; fitting the quarry spec red and white chevron vinyls, applied to the massive rear counterbalance, and the fire suppression system. The fire suppression fitters are installing a heat sensing pipe – rated to burst at 130 degrees – to the hydraulic pump and engine bays. Each of the two 9kg fire extinguisher units has the capacity to flood – through spray nozzles – each compartment with around 90 cu.m of dry powder.

Saturday, 20 July. Kevin's team spends most of the morning power washing the machine, removing any slight marks from the paintwork and spray painting any areas that had bare metal or exposed bolt heads. The cab interior receives a full valet too, as the team intends to hand the machine over as if it had rolled off a Komatsu factory production line.

Sunday, 21 July. Rest day.

Monday, 22 July–Wednesday, 23 July. Final adjustments are made to the boom and stick by adding a few extra shims here and there at the pivot points. Since the PC2000 is fitted with an extensive autolube system and 45 gallon drum of grease, it's a long and laborious job to bleed all the grease points and ensure the system works well. At the back of the field assembly manual there is a comprehensive 13-page check list that must be completed, and each section is signed off by Kevin; tick boxes are used, and pressure readings taken and recorded, along with oil levels. It is such a comprehensive list, just about every part that was fitted gets rechecked just to make sure nothing was missed or left loose, so as you can imagine, that takes up all of Kevin's time on Wednesday.

Thursday, 23 July. At 9am Kevin hands over the keys of the PC2000 to his colleague Peter Barrett, who will carry out production testing and operator familiarisation training for the quarry's shovel operators over the next two days. Kevin remains on site during this process, just in case he's needed to fix any minor issues. Having sold the first PC2000-8 FS machine into Europe, Koji Ito, European product support engineer has arrived

Kevin is in charge of quality control and checks the side panel 'shut lines' for an accurate fit and finish.

Track motor covers being fitted. Just about every big part of this monster machine is a heavy lift, requiring the use of Brian's crane.

At the end of two weeks' hard graft, the finishing touches are being completed on time and budget.

The comprehensive build and quality control manual is used from start to finish.

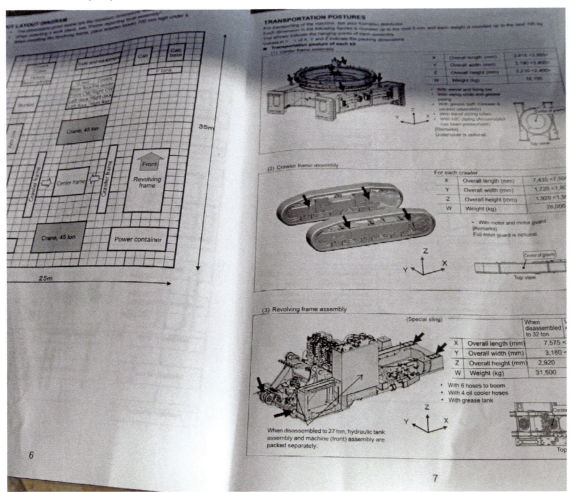

MKL has subcontracted only two areas of the build; fitting of the quarry spec red and white chevrons vinyls, applied to the massive rear counterbalance, and the fire suppression system.

from Belgium to see the machine in action and to offer Dave Hurcombe some advice on how to get the best production performance out of his new PC2000-8 face shovel.

Dave said: "Mr Koji Ito had some interesting advice to give; and the operator's first impressions of the PC2000-8 shovel compared to the old Demag H135S were that it's a very smooth, fast, quiet and powerful tool, for four or five pass loading our 91 tonne capacity trucks. Kevin and his team have worked safely and professionally throughout the build and have followed all the site rules, set up their own control procedures, documented all the risk assessments and, just like the last build, I'm very pleased the job was completed with no incidents. They are a pleasure to work with and a really good crew."

In the next chapter I make a return visit to see the PC2000-8 face shovel in action.

CHAPTER 20

Lafarge Tarmac • Komatsu PC2000-8FS operational visit • Dunbar • August 2013

Following on from the previous chapter, where I spent a number of days covering the PC2000-8 face shovel being meticulously assembled by a team of skilled fitters from Marubeni–Komatsu Ltd, I was looking forward to a return visit to see it working.

Dave Hurcombe, quarry manager, is delighted to have the new Komatsu PC2000-8 face shovel safely built on site and without incident. He is also relieved that the old Demag-inspired Komatsu H135S face shovel kept going while the order was being filled for a 200-tonne class PC2000-8 face shovel (FS), manufactured in and shipped from the Osaka plant in Japan.

Dave is also pleased to report that the load and haul package of Komatsu excavators and four Cat 777F 100-tonne capacity haul trucks has proven to be a good choice for the company. With the PC2000-8 backhoe still sipping fuel at an incredible 90 litres per hour, track pad wear is good and after 3,000 hours of hard digging, the machine has been ultra reliable and the backhoe bucket has just had a visit from the 'dentist' for a replacement set of Hensley XS145 teeth! Dave is expecting the same level of performance from the PC2000-8 face shovel.

With 10,000 tonnes of limestone — or overburden

Dunbar cement factory and quarry marks 50 years of operation during 2013. A new bench on the left and the ramp to the tip area on the right — this is a cut and fill operation.

– to shift each day, the PC2000-8 FS machine, fitted with its 11.5 cu.m bucket, will offer a significant performance advantage over the 135-tonne Komatsu H135S fitted with a 9.5 cu.m bucket.

Dave Hurcombe comments: "The main drivers for investing in a new 200-tonne class machine were simply that the old Demag shovel needed to be replaced, and in its 50th year the cement works' output has increased. These factors focused our attention on a state-of-the-art bigger machine to match our new truck and shovel operation and the increased demand for cement products.

"We looked at a number of competitor machines and having 700 hours' experience with the Komatsu backhoe, driver feedback, machine performance, service back-up and lower cost fuel burn, on balance the PC2000-8 was the best option for us and the deal with Babcock International means we have known fixed costs, which is good for our business.

It would be easy to unduly criticise the old H135S – in a number of areas – compared with the latest PC2000-8 FS machine. That said, for the operator, the reality of getting to and from the cab is simply night and day, as the PC2000-8 is fitted with a hydraulically operated drop down step arrangement that has double handrails, compared with the manually retracted and locked aluminium ladder set-up found on the old machine. Plus, there is also an escape ladder near the cab to shimmy down should the machine suffer an uncontrollable fire at the rear.

Safety is of paramount importance to Dave Hurcombe and the rest of his Lafarge Tarmac colleagues. With that in mind, Lafarge Tarmac specified other safety features as optional extras, such as the three rear-view cameras (one fitted to each corner and centre of the counterbalance), quarry spec warning beacons (the green one covers the driver's seatbelt use) and red and white chevron vinyls applied to the rear of the machine. Other visibility aids are the high intensity blue strobe lights on the rear of the machine that operate whenever the travel motors are turning.

At the time of my visit the PC2000 FS had only 120 hours on the clock and relief driver Gavin Strang was starting his third shift on the new machine. Gavin commented on its performance: "The machine is very stable in comparison with the old Demag as there is a considerable weight difference and it has a longer back ballast, which I just need to be mindful of when spotting the dump truck to ensure a safe swing radius."

From centre to rear ballast, the Demag is 4.73m long compared to 5.88m on the PC2000 and has a working reach of 11.5m vs 13.7m on the new face shovel, something Gavin has to take into consideration as he becomes accustomed to the differences between the two machines.

He continues: "Apart from the extra weight, comfort and power, the really big advantage of the new Komatsu is the height of the cab, as I can see right into the skip of the truck now, whereas before – operating the Demag – the driver's eye line was just in line with the top of the truck's skip. The overall visibility is excellent and it has a great rear-view mirror and camera package. With the face shovel machine it feels a bit easier and quicker to operate, as you're only putting the bucket into the face, lifting and dumping compared to the extra movements of the backhoe machine."

He adds: "As far as daily checks go, the machine has been well engineered as there are numerous lights in the engine and hydraulic pump bays, so we can check the levels day or night and the auto-grease system is really good, with a flag on the LCD screen when it kicks into operation."

The cab is fitted with a substantial rock guard across the large front window and Gavin commented that he doesn't even notice it's there any more. However, he and the regular driver, Andy McClung, feels that the travel levers slightly obstruct the view to the bucket more on the face shovel than they do on the backhoe version and so they will ask for them to be removed during its first service as the foot pedals will suffice.

The old Demag H135S has served Lafarge well over its 14 years at Dunbar and with 25,754 hours on the clock, it still appears to be structurally sound. The Japanese-manufactured PC2000-8 has some Demag DNA running through its 'hydraulic lines', as the front end equipment was designed by Komatsu Mining Germany (KMG) (at the former Demag factory in Düsseldorf). Given KMG's long and enviable reputation and design experience in large face shovel technology, this should bode well for future reliability. The drop down access step is also the same design as that found on the KMG-designed PC3000-6 mining shovel.

The end of an era, the last Komatsu Demag H135S to operate in the UK.

At the heart of the PC2000-8 beats a powerful 30.4-litre, 12-cylinder twin-turbo with after-cooling, producing an impressive 956hp from the Komatsu SAA12V140E-3 engine. This is the same engine that has been tried and tested in both the Komatsu 785 91-tonne capacity haul truck and its big D475A 113-tonne dozer. The PC2000-8 is fitted with Komatsu's Total Power Management system, which minimises power losses in the hydraulics, cooling fan and PTO to dramatically reduce fuel consumption. Based on the first four weeks of operation, the new PC2000-8 face shovel is consuming diesel at an average rate of only 90 litres per hour, which is very impressive for a machine of this size. That is the same fuel burn as the much smaller Demag, resulting in a significant cost saving due to substantially increased productivity from the big Komatsu face shovel.

Like its sister machine, the face shovel has very low levels of exterior noise for a machine in the 200-tonne class. This is attributed to the power module, which features good sound insulation and the latest technology in large sound absorbing fan blades that help to reduce noise.

During my visit the PC2000-8 face shovel, with its big 11.5 cu.m bucket, was excavating a 15m deep band of unblasted shale. The face shovel was loading the Cat 777F trucks in five quick passes, knocking two passes off the cycle times previously taken by the H135S. And when shifting limestone, it can achieve a full load on the 91-tonne capacity Cat 777F trucks with four big passes in around one minute forty seconds. Once loaded, the trucks made the short – 500 to 1,000m – journey to the tip site where a Cat D7N was on hand to deal with the material. When hauling limestone, the distance to the primary crusher was only another 100m further on.

Once the excavators reach the hard limestone, the quarry's strategy is to blast to load and achieve high fragmentation of the rock to keep the 1,250 tonne per hour, primary crusher operating at peak efficiency. That said, the new PC2000-8 can use all its digging force of 721kN (73.5 tonnes) – significantly more than the H135S 600kN and slightly ahead of the PC2000-8 backhoe's grunt – for fast cycle times.

Operating a face shovel gives Dave the flexibility – over the backhoe – to work with higher limestone

The PC2000-8 face shovel is loading the Cat 777F trucks in four or five good passes.

Tipping the scales to 200 tonnes, the PC2000-8 is a stable heavy duty shovel.

A good view into the dig area and instead of foot pedals found on the H135S, the button on top of the right-hand joystick is used to open the bucket visor, the left joystick button to close it.

The drop down fuel coupling makes it easy for a quick pit stop to grab the 1,300 litres of fuel per shift.

A large, roomy operator's cab, which is taken from the 300-tonne PC3000-6 parts shelf.

Well-designed steps for safe access up to the upper structure and cab.

benches and it has drop-ball capabilities, too. Both excavators are normally teamed with two trucks each, as they have to blend two types of limestone together to produce the final product.

Since the quarry's primary crusher can only handle the hard limestone rocks that are no bigger than a square metre in size, Gavin and the other skilled drivers of the PC2000-8 FS use a giant drop-ball to break up the larger pieces of stone, which can weight up to 20 tonnes. At the time of our visit this technique had not been carried out using the new face shovel. It is worth mentioning at this point that the Demag has foot-operated controls to open and close the bucket visor and the PC2000-8 is fitted with buttons situated on top of the European spec short throw control levers.

Gavin and the other operators were advised during their familiarisation training that the visor cylinders are very powerful, quick in operation and the bucket should ideally be closed during the swing cycle to increase efficiency and dampen the hydraulic flow. So with that in mind, Gavin

initially thought that picking up and using the drop-ball was going to be a challenge, but as it turned out, he quickly got to grips with the ball and found it easier than expected. He was still able to hit the target every time with a precise release of the big steel missile, shattering the blocks of stone into smaller pieces in one powerful strike.

Dunbar quarry invested around £5 million in new front line equipment in 2012. The Komatsu PC2000-8 face shovel added about another £2.5 million to the total spend and was part a much bigger deal with Babcock, as covered in previous chapters.

At the time of my visit the Demag H135S was parked up awaiting disposal and will be the last Demag H135S machine to be operated on these shores, as the other four machines that came to the UK have either been exported or are believed to be scrapped. However, as one chapter closes, a new chapter begins, with Europe's only PC2000-8 face shovel starting to keep Scotland and the North of England supplied with cement products for many years to come.

Target practice with the PC2000-8 face shovel, which was easier than expected, using a massive steel ball!

CHAPTER 21

Wm Thompson & Son • Cat 336E H Hybrid excavator • Dumbarton • November 2013

Wm Thompson and Son was established in 1946 as a coal merchant in Dumbarton and is entering its fourth generation as a family-run business, with established links across the west of Scotland. The company is a significant player in quarry products, with clients ranging from private individuals and local builders to national construction and infrastructure companies.

Its Sheephill Quarry is situated just off the Dunglass roundabout on the A82 at Bowling, with good access to the west coast and only a few miles from Glasgow city centre, or south over the River Clyde via the Erskine Bridge to reach its customers. It has been providing quarried whinstone, a form of basalt, for more than

40 years and the site has future reserves and planning consent to continue for more than 30 years. It is able to offer a one-stop-shop with a large fleet of modern Scania HGV vehicles, allowing it to quickly and efficiently deliver quarried products to sites, or collect and transport inert materials to its licensed landfill site.

Wm Thompson & Son has been operating Caterpillar excavators exclusively at its quarry throughout this period and started out with machines such as Cat's legendary 225 series excavators. When Caterpillar's revolutionary 336E H Hybrid excavator was unveiled in October 2012, at Caterpillar's proving grounds in Peoria, Illinois, Thompsons was interested in owning the first

The 43ha Sheephill quarry is capable of producing up to 750,000 tonnes of product per annum, anything from armour stone to whindust, and all sizes in between.

one in the UK, as Caterpillar tested the machine in 90 and 180 degree same-level truck loading, trenching and bench loading, and claimed operators could expect to use up to 25% less fuel compared with a standard 336E and to notch up a 33% improvement in fuel efficiency against the older 336D model.

During initial discussions for a standard 336E replacement machine, and using data from its current 336D fuel burn, Finning UK's Alistair Murdoch's bold challenge to the quarry firm was this: "Give us an extra £25,000 for the latest hybrid machine and we'll save you £70,000 over the next five years."

Not surprisingly, Andrew Thompson, operations director – and fourth generation of this long established family-run business – was initially taken aback by the additional £25,000 that he would be paying for the 336E hybrid machine. However, Finning presented the calculations that the hybrid system would save the business around £70,000 in fuel costs over five years and the benefits of ownership were quickly apparent as this hybrid solution would not just save money, but also improve productivity at the company's main quarry.

During my visit I asked Andrew about his decision to become the first UK customer for the 336E hybrid excavator and he replied: "We have a longstanding relationship with Finning and Caterpillar equipment and when our Cat 336D machine was due for replacement I spoke to Alistair Murdoch about the new Hybrid excavator. Since we operate in a competitive sector, and whilst some may hesitate at buying into new technology, we do not. I'm confident that Caterpillar have a good track record at launching innovative technology and they have a big R&D budget to get it right first time.

"This hybrid machine will save a significant amount of fuel which will help us remain competitive and raise productivity going forward. When buying new kit, we are always on the lookout for a machine that will add value or give us an advantage in our business and the 336E H does that, with the latest engine technology and hybrid system, coupled to excellent reliability, superior levels of comfort. And our operators are delighted to be in the seat of a Cat machine."

One of the other buying considerations is the good level of service provided and response times from Finning when it needs to resolve technical issues quickly to minimise down time. Thompsons has its own fitters to maintain its large fleet of HGVs in-house and carry out all the routine servicing itself as the Cat machines are easy to work on, with good access to fuel and oil filters, and Andrew rates Finning's parts service as excellent. Andrew did not take out a maintenance contract with this machine, however, he recognises that a time might come when – with technology becoming more complicated – Cat engineers will carry out the servicing. Having said that, Thompson does have a service contract with Finning for all its Cat-powered crushers and screeners at the quarry.

Before operating a number of Cat 336 size machines, the quarry operated two 330D models and during a large expansion phase a few years ago they were working on higher benches, casting off material, racking up high hours tracking around the site and sustaining minor body damage due to working in tight pockets. Despite this, the residual values on his Cat machines still remained high, which helped to persuade Andrew to remain with the Caterpillar brand.

The previous Cat 330 and 336 models were fitted with standard undercarriage and track pads. While this was suitable for stripping off softer material, the new 336E H will spend its days on the hard quarry floor, and with that in mind it was ordered with a heavy duty undercarriage, narrow 600mm double grouser track pads and full length track guides. This set-up will help to eliminate damaged and broken track pads caused by travelling over the large rocks and the operator has already noticed a significant improvement in stability and grip with this specification.

At the business end of the machine, the 336E H was specified a direct mount 66in, 2.19 cu.m bucket made by Millar, a MB700 model, which is designed for aggressive digging and loading of highly abrasive materials such as granite and basalt rock. Additional equipment includes a full length rock guard across the cab front screen and roof window, as well as quarry spec seatbelt green warning light. The 336E H was ordered with a heavy duty 6.5m (21ft 4in) heavy duty reach boom and a 3.2m (10ft 6in) long stick, which gives it the necessary breakout force of 166kN.

During my visit, the 336E H only had 105 hours on the clock and I found it working hard using nearly all its 7,490mm reach to dig out a trench of near solid rock to create a drainage channel, as heavy rain had raised the water level in the lowest part of the quarry floor. The water had to be drained before the 336E H could tackle the newly blasted face containing more than 50,000 tonnes of rock. Despite the hard digging, the hybrid machine is currently sipping fuel at 8 litres per hour less than the previous 336D L machine. Finning is confident that once the excavator gets back to its normal duties of bench loading the heavily blasted material into one of the two primary crusher units it will comfortably deliver more than a 25% fuel saving over the previous machine.

And in terms of CO_2 reduction, put into context, Cat claims that a 25% fuel advantage is equivalent to removing 12 cars off the road annually and in financial terms, with the current price of diesel fuel the 336E H is saving about £50 per day or £12,000 per annum. As the machine moves back to its normal, less arduous, loading duties, the £25,000 extra investment over a standard 336E will be recouped in only two years' time.

Andrew Thompson is able to keep a close eye on fuel burn via Caterpillar's GPS system, Product Link, and he can also check idle time, as well as machine utilisation. Additional fuel saving features start with an auto engine shutdown, which has been set by Finning to kick in after three minutes.

At the heart of the 336E H hybrid system are two large nitrogen-filled hydraulic accumulators, stationed in the counterweight, which are designed to harvest energy during swing breaking phase – at the end of the swing cycle – and then unleash the stored energy back into the hydraulic swing motor when the 22-tonne upper structure needs to accelerate back to the dig position. This system helps to produce better fuel burn than the standard 336E model. The energy released from the accumulators is reported to be about 150hp, which is about 46% of engine horsepower. From a simple engineering point of view, it is an inspired design to use hydraulic accumulator technology to create a hybrid machine.

The other three main components to make the hybrid system work are the Adaptive Control System (ACS) valve, Swing Energy Recovery (SER) valve, and the ESP (Electronic Standardised Programmable) pump. ACS is the electronic brain of the hybrid system and, with its 16 metering valves, it manages hydraulic flow to improve fuel efficiency and controllability.

The SER valve controls the capture and releasing of energy in and out of the swing accumulator. The ESP pump has an electronically controlled swash plate and the pump senses load on the engine and increases or decreases power to meet demand during the whole dig cycle, which results in fast cycle times at lower engine speeds and improved efficiency.

The 336E H operator, Craig Cameron, was settling in nicely to his new 'office' and after two weeks of operation had some positive comments to make: "It's really quiet and comfortable, which is an improvement over the 336D model. Visibility out the cab and mirrors are also good and the rear-view camera system works well, with its graduated zone lines on the large LCD screen to help judge distance to objects.

"Power wise, I've never really had to set the machine in anything other than the eco mode when loading the crusher and despite the hard digging of recent days it's still in the eco setting. If you do need a bit of extra power, then it's easy to select standard or power mode, as it does make it bit quicker – it's definitely a nice bit of kit."

Craig has also found a number of the features beneficial, one being the warm-up mode at the start of his shift as this allows him to start the machine and in five minutes, by the time he has given the crusher a check over, the 336E Hybrid hydraulics are up to working temperature. On the 336E Hybrid model, one other feature that may be useful is the electronically selectable joystick control pattern – to suit driver preference – through the cab LCD screen and monitor.

Both Andrew and Craig praise the levels of comfort found in Cat machines and inside the spacious ROPS cab of the 336E H is a comfortable heated and air suspension seat, which also comes with height adjustable joystick levers, armrests and all the other modern creature comforts. The 336E H is quiet on the outside as well as having a claimed rating of only 65.2dB(A) of sound inside the cab.

Purring quietly away at the back of the 336E H is a Cat 9.3-litre six-cylinder Stage IIIB turbocharged and after-cooled engine, (the same power unit fitted to the D6T dozer) producing 328hp (100hp more than the D6T) at a very low 1500rpm. Like all Stage IIIB Cat engines, the 336E H is fitted with an EGR system that captures and cools a small quantity of exhaust gas, then

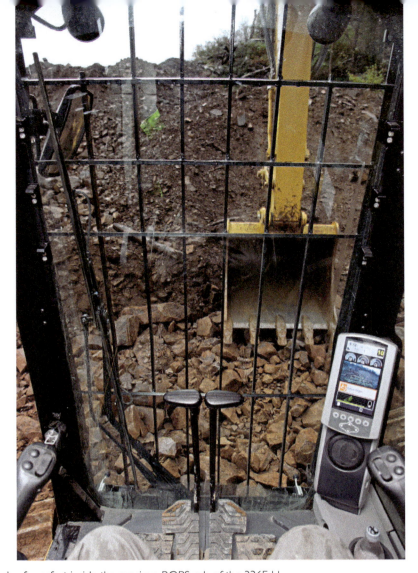

The operator praises the levels of comfort inside the spacious ROPS cab of the 336E H.

The office with a view over the River Clyde, Erskine Bridge and Glasgow city.

A Millar direct mount MB700 bucket was ordered, which is designed for aggressive digging and loading of highly abrasive materials such as granite and basalt rock.

Cat 336D L excavator was loading the other primary crusher with blasted stone.

Making the hybrid system work are the Adaptive Control System, seen here next to the boom foot, and the Electronic Standardised Programmable pump, seen at the o/s/r of the machine.

Cat 972 H wheel loader stockpiling material from the primary crusher.

routes it into the engine combustion chamber where it drives down combustion temperatures and reduces NOx emissions.

In addition to this, to meet EU Stage IIIB emission standards, after-treatment components include a Diesel Oxidation Catalyst (DOC). This uses a chemical process to convert regulated emissions in the exhaust system, and a Diesel Particulate Filter (DPF) to trap particulate matter (soot) that is carried into the exhaust gas stream after combustion; Cat's regeneration system is designed to work without any interaction from the driver. A soot level monitor and regeneration indicator lights are integrated into the 336E H LCD screen.

Sheephill quarry also operates a number of other Cat machines and the 43ha site is capable of producing up to 700,000 tonnes of product per annum, anything from armour stone to whindust, and all sizes in between. Working with the 336E H on the quarry floor was a 320D L excavator and a Cat 972 H wheel loader was stockpiling material from the screeners and also loading a steady stream of Thompsons' six- and eight-wheeled tipper HGV trucks. Meanwhile, up at the top of the quarry, working a seam of recently blasted stone, a Cat 336D L excavator was loading the other primary crusher and was teamed with the other Cat 972 H wheel loader on similar duties.

Since Andrew is willing to invest in the latest technology to drive efficiency and lower costs, I return to this site in Chapter 23 to look at a Cat 972M XE wheel loader fitted with Caterpillar's new CVT transmission and find out how the 336E hybrid is performing over a longer term of operation.

At the heart of the 336E H hybrid system are two large nitrogen filled hydraulic accumulators (top left) stationed in the counterweight, which are designed to harvest energy during swing breaking phase. The Electronic Standardised Programmable Pump (centre) and Adaptive Control System Valve (far right) are the other main components. Photo: Caterpillar Inc.

CHAPTER 22

Cloburn Quarry • Cat 966K XE wheel loader • Lanark • July 2014

Cloburn quarry is situated just outside Lanark – on the A73 – a mile south of Hyndford Bridge, South Lanarkshire – in the beautiful Scottish countryside and to the passer-by it is virtually undetectable from the roadside as the quarry is situated at the top of a hill, with the main plant and buildings recessed into the hillside.

The quarry's distinctive red granite stone was first extracted commercially from this location in 1896 to provide track ballast for a string of new railway lines. The quarry was first operated by Lanark County Council – along with some other smaller quarries – and named Cairngryffe after the hill that formed the deposit of red granite. In 1935 the council closed a number of smaller quarries nearby and installed what was then a state of the art crushing, screening and asphalt coating plant at Cairngryffe. This is why many of the roads in Lanarkshire were bright red in colour.

For the first 90 years, Cairngryffe operated as a local authority quarry but in 1986 it was taken over by the current operating company, Cloburn Quarry Co. Ltd, who replaced Tilcon at Cloburn quarry, located on the other side of Cairngryffe Hill. Today the two have been joined to become one of the UK's largest mainland quarries, supplying high quality granite aggregates to a wide range of customers throughout the UK, Europe and the Far East. An extension to the hard rock quarry has been granted recently to secure its future well into the twenty-first century.

Cloburn Quarry management looks to upgrade its machines on a five-year cycle, and with two wheel loaders from Komatsu and Daewoo due for replacement in late 2013, Colburn quarry manager Jim Erskine started to evaluate options from a number of manufacturers. With a Caterpillar 980H and a Cat 966H already in the fleet, Jim spoke to Alistair Murdoch, Finning UK's key account manager, about its Just Add Diesel offer and the benefits of the new Cat 966K XE wheel loader fitted with an advanced Continuously Variable Transmission (CVT). Following discussions, Finning UK delivered a 966K XE demo machine along with operator training, which was carried out by John Blackett, one of Finning's skilled machine demonstrators and trainers.

During the 966K XE demo period, competitor comparison tests revealed a very impressive 35% fuel

Cloburn Quarry is situated in the beautiful Scottish countryside and to the passer-by it is virtually undetectable from the roadside.

Approximately 230,000 tonnes of new blasted stone hits the quarry floor for the Cat 980H to deal with.

saving opportunity, which resulted in management placing an order for a 24-tonne Cat 966K XE machine. By doing so, Cloburn Quarry became the first company in Scotland to operate this model of wheel loader. At the same time, the management team also invested in the slightly bigger 26-tonne Cat 972K wheel loader, with both machines purchased using Finning's fixed costs and guaranteed uptime Just Add Diesel solution.

Jim Erskine, comments on its new 966K XE: "We have a large site, with a high rate of production, so we opted for a Cat 966K XE as we knew the payback for the additional investment for the XE model would be achieved quickly. Finning provided us with a demonstrator machine and we carried out performance comparisons against other manufacturers and found fuel savings of 35% being achieved. We selected the Cat 972K because it meets our requirements for a high-performing, durable and versatile machine. And the Just Add Diesel option on both the 966K and 972K means

we have known fixed costs with a total service package from Finning and guaranteed machine availability to maintain productivity."

Jim adds; "Finning's parts and service back-up is first class, so this gives me confidence that they are in good shape to keep their guaranteed uptime promise. Quarrying demands machinery that is durable and robust, and with the added benefit of the Finsight engineers constantly monitoring the wheel loader's performance, I am sure we will see benefits to the business over the next five years."

The Just Add Diesel deal from Finning guarantees its customers 98% machine uptime and fixed costs, while the machines benefit from the Finsight condition-monitoring tool, a five-year/10,000 hours extended warranty, and a five-year preventative maintenance package. This allows Cloburn Quarry to manage expenditure and reduce the risk of escalating service costs. Apart from wear on tyres and ground engaging parts over the term of operations, all the operator is responsible for is simply filling the machines with fuel.

At the heart of the Cat 966K and 972K is Caterpillar's C9.3 9.3-litre engine, which has has six cylinders and is electronically controlled with turbocharging and after-cooling. It uses a small amount – less than 18% – of Exhaust Gas Recirculation technology and after-treatment components include a Diesel Oxidation Catalyst and a Diesel Particulate Filter (DPF), which allows the machine to meet Tier 4 Interim/Stage IIIB emission standards. Caterpillar claims 330 gross hp for the XE machine, which is 8hp more than the standard 966K while spinning at 100rpm lower, at 1700rpm! And it produces 1,490Nm of gross torque at a very impressive 1200rpm vs 1,426Nm at 1400rpm on the standard 966k. This engine technology coupled with the CVT make for a very efficient powertrain to maximise fuel efficiency and this helps Cloburn Quarry to produce material at the lowest cost per tonne moved.

While CVT technology is advanced for wheeled loaders – due to them mainly operating over short distance cycles between the pile and the loading point – this technology has been used in agricultural tractors for more than 20 years. This has not gone unnoticed by Caterpillar, as it has been developing and testing this type of transmission extensively for a number of years before bringing it to market.

The CVT gearbox works by having both a mechanical drive path – directly connected to the engine flywheel – and using a variator control pump connected to the hydrostatic drive motor, which is driven off a spur gear from the main mechanical driveshaft. At the other end of the mechanical driveshaft is a planetary gear set, which has an outer ring and this ring is connected to the hydrostatic drive motor.

When the 966K XE is stationary, both the main driveshaft and the outer gear ring turn at the same speed, cancelling each other out. In order to pick up drive the outer ring slows down to create drive and will eventually come to a complete halt as the speed of the wheel loader increases – at this point, only the mechanical drive is delivering the power. The outer ring will actually start to rotate in the opposite direction to the mechanical driveshaft to further increase speed and add additional power again. The hydraulic variator takes the place of the power-sapping torque converter and

has reduced heat generation under severe rimpull load. Caterpillar claims that when aggressively digging, the continuously variable transmission consumes roughly half the energy of a conventional transmission!

With a CVT system, in essence, there are no fixed gear ratios, just a constant variance in drive. For a simple explanation of how it works, Caterpillar has produced a short animation of the transmission in action, which can be viewed on YouTube at http://youtu.be/pD-I9CWfCfg

From an operator's perspective, controlling the power and transmission is by a simple rocker switch on the steering joystick and a two-pedal arrangement (three pedals on the standard 966K) that Caterpillar simply describes as the stop pedal (on the left) and the go pedal (on the right). The left pedal has three stages: to modulate the rimpull (torque), provide engine braking/retardation and when fully depressed it applies the service brakes. The right pedal controls the engine throttle and transmission pump pressure. The driver can also select a speed range in one tenth increments from the switch mounted on the steering control; this is useful when the top speed needs to be limited, otherwise the driver selects the highest 'gear' speed and the CVT does the rest. This leaves the driver to concentrate on the job in hand – not shifting gears.

Barry Wilson has been operating wheel loaders at Cloburn Quarry for more than eight years and was a big fan of the departed competitor machines – that was until the 966K XE demonstrator arrived. After receiving training from Finning, to get the maximum out of the XE machine, and after adjusting to the two-pedal operation, Barry has nothing but praise for the 966K XE. He comments: "The 966K XE is a great machine to drive as its very comfortable and has really good around visibility as the view out the large front window is faultless. Having got used to the joystick steering I would be gutted if I had to go back to a steering wheel arrangement. And I feel this wheel loader has so much more power, speed and fast hydraulics – I wouldn't want to drive anything other than this Cat machine."

Now that Barry has 1,200 hours of operation under his belt on the 966K XE, he also mentioned that with the correct combination of left and right pedal use, fast loading of trucks can be achieved with only the slightest of dab of the service brake needed to stop the machine. Clearly this will translate into lower operating costs with

reduced wear on brake packs and heat, so much so, that Caterpillar has removed the need to fit oil coolers to the axles.

The Cat K series wheel loaders have a clamshell two-piece engine bonnet design, with the rear half lifting on gas-filled struts to allow easy access for drivers to perform daily checks, plus the full bonnet is hydraulically powered to allow access to the powertrain components.

The two new wheel loaders will work alongside the quarry's prime mover, a Cat 980H, fitted with Rud tyre chains. Jim is also delighted with the performance of this machine, and the tyre chains have proven to last more than 11,000 hours, leaving the original tyres with 85% of remaining tread life after 10,000 hours of operation. The chains are needed in this application due to the abrasive nature of the granite when carrying out loading duties at the quarry face. The two new K series machines, a Cat 966H, two Komatsu WA470 loaders and a WA500 are tasked with loading and stockpiling duties.

Cloburn Quarry only has one ADT in its fleet, a Volvo A30E, and finds it more cost-effective and flexible to hire in the remaining ADTs. Some of these trucks are fitted with special treadless rock tyres on the rearmost axle and are normally teamed with the 980H wheel loader and used to transport the newly quarried stone to the primary crusher, however at the time of our visit it was being stockpiled due to planned maintenance on the crusher. Jim explained that the quarry is capable of producing more than 50 different products and that it keeps sufficient stocks of its main product lines, including a significant number of bulk bags of red granite chips; and can produce +68 PSV aggregates.

Quality is taken seriously and therefore to minimise any possible contamination, product is produced to order by the quarry's extensive array of secondary crushers and screeners to match customer demand. The instantly recognisable red granite is produced for a variety of domestic and commercial applications and is supplied throughout mainland Europe as well as for the USA and the Far East. At the time of my visit, the 966K XE was busy loading a constant stream of 44-tonne HGV articulated vehicles of a local haulier, Tennant Transport of Forth, that were heading to the Rosyth docks to fulfil export orders.

The 966K XE has an on-board weighing system fitted as standard, however, Jim chose to fit all his wheel loaders with on-board weighing systems. The 966K XE has the Loadmaster 8000iX, from RDS Technology, which is designed to weigh the load accurately to ensure the HGV tippers are loading within the legal limit for on-highway use and the quarry has a weighbridge to double check and invoice when supplying external customers. The on-board weighing system works by measuring hydraulic pressure in the lift cycle and weighing scales give a display in the cab of each bucket load and the total material loaded. That said, modern HGVs now have their own on-board weighing systems, and if fitted the operators will take hand signals from the truck drivers as they start to tip the last bucket load into the body.

Jim also chooses to order bespoke buckets from either MST or Miller, (both are based in the north-east of England) and are constructed from thicker Hardox steel than standard to cope with the abrasive granite material. The bucket size is also increased, up from 4.2 cu.m to 4.8 cu.m on the 966K, which makes for efficient loading in four easy passes.

Cloburn Quarry has been supplying high quality track ballast to the railway sector for 118 years, and during my visit was no exception, as I observed the larger 26-tonne, 322hp, 972K, fitted with a 6 cu.m bucket loading an eight-wheel tipper with this material in just over two passes!

Graham Patterson is the driver of the 972K and comments on its performance: "The 972K is a good tool, far better than anything else I've driven before. A very comfortable machine to spend shift in and I feel less tired using the joystick steering compared to any of the other wheel loaders here fitted with a steering wheel."

To keep tyre costs low, Michelin's XLDD2A tyres (used instead of standard tyres) are fitted to the wheel loaders and claim to offer an aggressive and open tread centre that delivers excellent tractive performance and an extra deep tread pattern that promises to give outstanding wear life potential. Jim is expecting more than 10,000 hours from them and in addition to this, since they have 40mm of extra tread depth, this helps the smaller 966K with a bit of extra pin height when loading high-sided trucks.

Finning product link data shows the 966K XE burning

Rud protective tyre chains have outlasted the machine before being traded in – as the tyres were found to be only 15% worn after 10,000 hours of use.

With the correct combination of left and right pedal use, fast loading of trucks can be achieved with only the slightest of dab of the service brake needed to stop the machine.

Volvo A35F and Liebherr ADTs are hired in and some of the ADTs have rock spec tyres fitted for this application.

A great view out of the pillarless front screen on the Cat 966K and 972K.

Operator Barry Wilson carrying out his daily checks from ground level.

Cat 966H and the newer 966K XE parked together. Note the sight difference in bucket design between MST and Miller.

an average of 16.1 litres an hour, and compared to Cloburn's 966H, which is burning 19.3 litres an hour, it has a 3.2 litre an hour advantage. At 2,000 hours per annum the 966K XE will save 6,400 litres of fuel, which is equivalent to 17 tanks of fuel per annum or, put another way, 32,000 litres (that's one full-sized road tanker)

over a five-year period! Finning claims that, due to the sheer size of the Cloburn site and the fact it lifts a bigger bucket, a typical 966K XE is capable of producing fuel burn as low as 12 litres an hour.

The 966K XE and 972K have been fitted with the usual array of quarry spec safety devices, such as radar

Cat 972 fitted with a 6 cu.m bucket and with just two big passes an HGV 8 wheeler is fully loaded with railway ballast.

warning equipment and blue flashing strobes, and the 966K XE has been given the red and white chevron treatment on the rear end.

With two very happy operators, a 35% fuel saving against competitor machines and with Caterpillar launching its M series wheel loaders with its ultra-fuel efficient CVT technology being made available on the Cat 972M model, Cloburn Quarry management has even more fuel efficiency options to choose from going forward.

Full details of the Cat 972M with CVT technology is covered in detail in the next chapter.

CHAPTER 23

Wm Thompson & Son • Caterpillar 972M XE • Dumbarton • December 2014

If you have read Chapter 21 on Wm Thompson & Son you will know they are no stranger to investing in the latest Caterpillar technology. They took delivery of the first 36-tonne 336EL Hybrid excavator in the UK during October 2013, and this uses a unique energy recovery system that operates during the machine's swing cycle to offer a significant reduction in fuel burn. Thompsons has since completed another first, by investing in a Caterpillar 972M XE wheel loader, the first one to be sold in the UK.

Finning UK claims the 25-tonne Cat 972M XE wheel loader is the class leader in fuel efficiency, with XE advanced CVT powertrain technologies enabling a 25% fuel saving for operators by using an integrated Cat Continuously Variable Transmission (CVT) system, rather than a standard, less efficient, torque converter. This means that the functioning of the XE model runs at 500rpm less than the standard Cat 972M. This will save an average of 4 litres of fuel for every hour the Cat 972M XE model is in use, and operating at 2,000 hours per annum this translates into a £4,000 saving on fuel costs.

Andrew Thompson, company director, said: "We have a long-established relationship with Finning and Cat equipment and when our five-year-old Cat 972H was due for replacement we spoke to Finning about the benefits of the new 972M XE model and then placed an order immediately.

"We constantly strive to be at the cutting edge of new technologies enabling us to maximise productivity and minimise cost, whilst operating in the most environmentally friendly and sustainable manner possible and we have every confidence that Caterpillar's latest technology will be nothing but reliable."

Andrew continues: "The new Cat machines are leading the way in our industry, therefore it was a simple business decision to integrate them into our fleet and benefit from the value they will add to our operation. The 336E hybrid machine has proven to be very reliable and saved us a significant amount of fuel, and I fully expect the 972M XE will be just as fuel efficient, which is good for the environment and our bottom line."

Wm Thompson staff have an opinion on the fleet of machines operating at the quarry, which Andrew will take into account. He said: "Making sure our operators know we're providing the best equipment possible is important to us. Operator comfort and machine functionality is a big factor in deciding to invest in the Cat machines. We've already seen a decrease in minor denting and uplift in productivity since using the Cat 336 Hybrid Excavator on site this year."

Operating the 336E Hybrid since its arrival has been Craig Cameron. Due to Craig's skill, care and attention, the 336E H – with more than 2,100 hours now on the clock – still looks like new. Craig mentioned that the 336E's performance has improved over the last year as the machine has 'freed off' to become even more productive. Andrew commented that he is delighted with the performance of the 336E excavator as it has not missed a beat and, despite the harsh rocky environment, the heavy duty 600mm wide tracks have required no adjustment. Andrew is also pleased with his team at Sheephill, as he can see that his investments are being well looked after and pointed to its seven-year-old Kleemann crusher as a good example of this; a typically unloved piece of equipment with next to no damage sustained over the years.

The 972M XE order was placed during February 2014 and the machine arrived on-site during the first week of December, which meant Thompsons kept its 972H

Updated from Chapter 21, the 336E hybrid is now loading the crusher where it belongs.

The 972M XE on stockpiling duties.

With its CVT transmission, the 972M XE is an ideal machine for load and carry duties around the site.

The two fuel efficient machines working as a team loading the crusher and HGV trucks.

The driver, Peter Leitch, says: "The all-round visibility is excellent and I really like the new joystick steering. It's a lot quicker and more responsive than the 972H steering system."

The standard Cat rear-view LCD screen and PreView system uses a patented pulsed radar technology to detect moving and stationary objects in the driver's blind spot.

The 972M engine is fitted with a new clean emissions module, and has more tech than a chemical factory!

The 972M working at one of the many stockpiles.

machine a bit longer than its five-year replacement programme. The new Cat 972M XE is operated by Peter Leitch, who has been an operator with Thompson since 1993 and has been driving wheeled loaders for the last 13 years at Sheephill Quarry. Peter's last machine was nearly six years old, a Cat 972H – which the 972M XE replaced – with 15,500 hours on the clock.

Commenting on the new machine, Peter said: "I really like this new shovel as it's a lot quicker at loading and carrying with the CVT transmission, which is absolutely seamless and there's no downshifting when you go into the stockpile now. When driving to the stockpile area, this 972M XE has a significant speed advantage over any of the other wheel loaders here, and compared to the 972H, even when fully loaded, it is at least 4kph faster and can easily reach 14kph up the haul road ramp."

Peter is not a fan of conventional steering wheel set-ups and his 972H was specified with Caterpillar's Command Control Steering, which looks like a steering wheel but full machine articulation is achieved with only a ±70 degree turn of the wheel, compared with two to three 360 degree turns of a conventional steering wheel. And it also contains the forward/neutral/reverse switch and the upshift/downshift buttons. Following on from the K-series, the new M-series loaders retain joystick steering, which is a low-effort electro-hydraulic joystick steering system fitted as standard. Cat claims its joystick steering system has an exclusive force-feedback feature that automatically increases joystick effort as ground speed increases. This improves steering control and comfort, especially at higher speeds.

A significant reduction in operator fatigue is achieved with this steering system as Finning claims that in general operation, a wheel loader fitted with a steering wheel, an operator would complete about 360 full turns of the steering wheel in one hour of operation – based on 40 minutes truck loading and 20 minutes travel time – and with the steering wheel being approximately 350mm in diameter, the operator would need to move a full arm length up to his shoulder, whereas, the joystick tilts 40 degrees side-to-side and the operator's arm is situated comfortably on the armrest and just needs to rotate the wrist or forearm to steer the machine. And for additional ease of operation the forward/neutral/reverse switches are incorporated into the handle.

The ease of the steering operation and the easy to

use CVT transmission switches and two pedal controls mean the operator of this machine can spend more time concentrating on the job in hand – not changing gear. And to further improve operator comfort and efficiency, the 972M model is now fitted with a single joystick to operate the loader's arms and bucket, replacing the two levers found on the K-series. Rocker switches can also be built on to the joystick to set the detents for the dump, hold and rackback kickouts.

The 972M was shipped with a 4.6 cu.m performance series bucket, which is loading an eight-wheeler in three passes or a six-wheel truck in two passes. That said, Andrew ordered a 7 cu.m – general purpose bucket to provide good all-around performance for stockpiling, rehandling, excavating and bank loading. Andrew took this decision as one of the 972H wheel loaders is already sporting a 6.8 cu.m bucket with no problems encountered. Once the 7 cu.m bucket arrives, Andrew is looking for the 972M XE to load an eight-wheeler in just two passes and the smaller six-wheeler in just one pass!

As mentioned in previous chapters, quarry operators I have visited seem to favour Michelin's XLDD2A tyres over the standard XHA2 tyres and this hard wearing type has been fitted to the 972M. Also helping with tyre wear is (apart from a skilled operator using the Go pedal correctly) Cat's new on-the-go, disc-type differential locks, which will help to improve traction on slippery terrain and are fitted as standard on the front axle. The diff-lock is manually activated by a switch on the floor or on the right-hand one-piece implement joystick and there is an option to fit fully automatic front and rear axle differential locks. The M class wheel loaders have new external caliper disc parking brakes mounted to the input shaft of the front axle for extra efficiency over enclosed wet systems and ease of accessibility for inspection and service.

With safety in mind, Andrew ordered the 972M XE with blind spot detection equipment by Preco Electronics. The PreView system uses a patented pulsed radar technology to detect moving and stationary objects in the driver's blind spot, and the makers claim it was designed for arduous environments. The driver is alerted by a variable audio sound and LED lights – the more lights and the higher the pitch of sound, the closer you are to an object. This has been fitted on, and to complement, the large LCD rear-view monitor fitted as standard. Other additional equipment is a set of rear-mounted blue flashing strobes, a seatbelt warning beacon and rear chevrons that have still to be fitted by Finning.

Purring away in the back of the 972M XE is Cat's well-proven C9.3 9.3-litre ACERT engine, the same power unit fitted to the D6T dozer and 336E H excavator. On this machine it produces 315hp (23hp up on the 972K) at only 1600rpm and peak net torque of 1,618Nm (which is a whopping 283Nm more than the 972K), which is achieved at an incredibly low 1200rpm.

In order to meet the latest Stage IV emission standards – which meant achieving an additional 80% reduction in NOx emissions over Stage IIIB standards – the C9.3 six-cylinder, turbocharged and electronically controlled engine is fitted with a new clean emissions module, comprising of five elements. Two of these, the diesel oxidation catalyst (DOC) that uses a chemical process to convert regulated emissions in the exhaust system, and a diesel particulate filter (DPF) to trap particulate matter (soot), can be found on the Stage IIIB engine in the 972K. The three additional systems – on the 972M XE – are the selective catalytic reduction (SCR) that consists of an SCR catalyst, ammonia oxidation catalyst (AMOX) and a pump electronics tank unit (PETU). This SCR design uses a small amount of Diesel Exhaust Fluid (DEF), aka AdBlue, to convert NOx emissions in the exhaust gas into nitrogen and water. Due to forward planning in Cat's design department, no major redesign was necessary to accommodate the extra SCR system components into the 972M XE rear engine hood, as the space required did not change from its Stage IIIB/Tier 4 Interim engine. A key component of Cat Tier 4 technology is injection timing, which precisely controls the fuel injection process through a series of carefully timed microbursts that provide more control of combustion for the cleanest, most efficient fuel burn.

The engine is also fitted with strata pre-cleaner system that removes 93% of the dust particles, using centrifugal force to spin dust and dirt to the outer walls where they are ejected out into the exhaust stream, resulting in extended filter life. The general maintenance of the 972M XE will be carried out in-house by Thompson's skilled fitters, with major servicing outsourced to Finning

UK. And with Finsight also included in the deal, Finning's support crew based in Glasgow will provide Andrew and his team with analytical support to help make sure its operation is running at its full potential, both in terms of productivity and efficiency.

Quarry staff are using intermediate bulk container (IBCs) to store the AdBlue fluid, which provides sufficient capacity to cover the SCR engine found on the Cat 972M XE, Doosan DL420-3 and DL450-3 Scania-powered SCR Stage IIIB compliant wheel loaders and some of their crushing equipment.

The Cat M-series loaders have viscous cab mounts to decrease the noise and vibration to which the operator is subjected, which helps to produce a very well-controlled 67dB(A) noise level in the cab of the 972M XE, and is 2dB(A) lower than the 972K. As mentioned across a number of different applications and machine types throughout this book, the cab is so well insulated, Peter struggles to hear the sound of the HGV truck horns and so has to keep the radio volume turned down low!

Sheephill quarry also operates a large number of other Cat machines on this 43ha site, which is capable of producing up to 700,000 tonnes of product per annum; anything from armour stone to whindust, and all sizes in between. At the top of the quarry, working a seam of recently blasted stone, a Cat 336D L excavator was loading the other primary crusher and teamed with a Cat 972H wheeled loader. And working with the 972M XE on the quarry floor was the other high profile Cat machine, the 336E H excavator. The 336E was loading a top of the range 2007 model Kleemann MC120 Z track-mounted jaw crusher, which has proven to be a robust and productive crusher unit.

The MC120 Z is another hybrid tool as the jaw is driven by its own diesel-powered generator, which feeds a powerful 215hp direct electric motor, providing high capacity crushing with low fuel consumption. Following Andrew's strategy of using the very latest fuel efficient technology, secondary crushing is carried out by a new Powerscreen 1300 Maxtrak cone crusher, which is capable of a 350-tonne per hour output. The Maxtrak is powered by a 12.7-litre Scania DC13 Stage IIIB engine, producing 450hp, which – like the Cat 972M – uses SCR (Selective Catalytic Reduction) technology to reduce the NOx content in the exhaust gases. And to complete the operation, the electrically driven stocking conveyer

The two fuel-saving Cats side by side.

is powered by an umbilical cord connected to the Kleemann MC 120 Z generator.

During my visit, the 972M XE was working at the stocking conveyer and loading a steady stream of Thompsons HGVs with type one material. It was also running this to the stockpile area in a lift and carry operation. Thompsons has also invested recently in a state of the art IT ordering system, which means front line quarry staff have instant access to the latest orders for material. Another duty of the 972M driver is to respond to these orders, which means moving the stockpiled material to feed one of the many screeners to match customer demand for different product sizes.

Thompsons is able to offer a one stop shop, with a large fleet of modern Scania HGV tippers powered by the latest fuel efficient Euro V and Euro VI compliant engines. With the Cat 336E hybrid at the quarry face, feeding two fuel efficient and environmentally friendly crushers, and the new Cat 972M XE at the other end of the operation – with each of the four machines saving 3 or 4 litres an hour – this must be one of the most fuel efficient and environmentally sustainable set-ups found in any Scottish or UK quarry.

Moving stockpiled material to feed one of the many screeners to match customer demand for different product sizes.

CHAPTER 24

Jim Jamieson Quarries Ltd • Liebherr R960SME • Ardlethen Quarry • July 2015

Jim Jamieson is the MD and owner of Jim Jamieson Quarries Ltd and has a longstanding relationship with Finning and the Caterpillar brand. He and his son, James, have a substantial fleet of Caterpillar equipment. However, after numerous meetings with Liebherr–Great Britain Ltd (Liebherr) and travelling to see a Liebherr–Rental's R970 in action, Jim has invested in a new 60-tonne Liebherr R960 in Super Mass Excavator (SME) specification – the first R960SME machine to be sold in the UK.

One of the reasons for Jim selecting the R960 is down to his experience with a pre-owned (2007) R954 machine bought shortly after he had taken back ownership of the 110-acre Ardlethen Quarry. The R954 now has 8,600 hours on the clock and has been ultra-reliable, as no one can recall it ever suffering a breakdown during its time at the quarry. This is a remarkable achievement, given that it is not a quarry spec machine, which bodes well for the new R960SME excavator going forward.

Jim also has a longstanding business relationship with a local self-employed fitter who carries out all planned maintenance, including oil sampling on his earthmoving equipment. However, Jim can also rely on the Liebherr engineers when called upon. Liebherr has plans to strengthen its presence in the north-east of Scotland by hiring additional engineers to service its expanding customer base and will be looking after the R960 during its first year of operation.

Prior to the arrival of the R960SME, the 50-tonne class R954 was the prime mover at Ardlethen and will remain at the quarry to help load the primary crushers, while the bigger 60-tonne newcomer has been brought in to carry out heavy duty rip and loading operations. The quarry is normally drilled and blasted to extract the stone, with high fragmentation for ease of loading, however, some areas of the quarry are starting to get close to some buildings and boundary fences, which will prohibit blasting activity.

This is where the Liebherr R960SME will take up the challenge of ripping out the hard rock and then switch to loading duties. The R960SME is a relatively new model in the Liebherr crawler excavator range, which replaces the R954, which is no longer available except in a demolition spec. The R960SME is an impressive looking machine, as Liebherr claims it has bigger and more powerful hydraulic cylinders on the boom, dipper, and bucket. The SME machine also has a stronger boom, with internal plating, and a choice of shorter dipper arms – 2.6m or 2.8m – for much higher breakout forces, with a 2.8m stick selected on this machine. However, what is really eye-catching is the undercarriage.

The R960SME is fitted with the undercarriage from the next size machine up in the range, the R966, and looking at the Liebherr spec sheet, the track width is actually wider, at 3.71m, than some 80 to 90-tonne class excavators and would not look out of place on a high-reach demolition machine. The undercarriage design, along with its heavy duty construction, should make the R960 very stable when ripping out unblasted rock. Jim has spec'ed the machine with 600mm wide, double grouser heavy duty track pads and sealed-for-life Liebherr D9K chains to cope with the arduous conditions.

The standard R960SME is well equipped for mass muck shifting duties. However, for quarry work, Jim has ordered a number of optional extras, such as the Integral protection guard (FGPS + FOPS) cab guard to protect the cab and driver. This will be particularly useful when ripping out material from the rock face, as the machine in SME spec (with its 2.8m stick) will have a 10m high

reach, when positioned 5m back from the face. The bucket/ripper hydraulic cylinder has a ram guard fitted to protect the cylinder rod from rock damage, while the heavy duty 6.7m long mono boom has been fitted with protection on the underside to match the protection plate fitted to the dipper arm as standard.

The other significant optional extra is the Liebherr HD77 quick hitch, which is made in the mining division next door to the main factory in Colmar, France (close to the German border), where the R960SME was manufactured. The quick hitch is a serious bit of kit, weighting in at 631kg and designed by Liebherr for quarry and mining applications.

The hitch has a conventional C section to engage with the front pin on the bucket or attachment, but unlike conventional quick hitches that you would find on smaller class excavators it has two large diameter hydraulic actuated pins that push out (at 150 bar pressure) sideways into the holes in the bucket or ripper attachment. This provides a rock solid connection and eliminates any play between the hitch and the bucket/ripper attachment. When the hitch is fully locked, this is displayed on the in-cab LCD screen.

The hitch is rated to operate at 27 tonnes, which matches the lifting capacity of the R960 (at a 3m reach)

and even at its maximum rated lifting capacity, the R960 can handle operating over the tracks or over the side at this weight! It makes no difference to the stability of the machine, due to the width and weight of the SME undercarriage design and large rear counterbalance weight.

Hi-Spec Manufacturing Ltd (HSM) has a worldwide reputation for making – and reconditioning – bespoke buckets for some of the largest mining excavators in the world. So when selecting a bucket and ripper for the R960, Jim arranged, through Liebherr, for HSM – a company within the MST parts group – to make the two bespoke attachments, a 3.3 cu.m rock bucket and a ripper tooth.

Jim likes the fact that HSM buckets are made from Hardox Swedish steel and chromium carbide is used throughout to give extra strength and reduced wear. He has also spec'ed the bucket with 450 Hardox plate throughout – including wear plates – and with ESCO teeth mounted on J600 adapters fitted with a locking pin system that is compatible with Caterpillar GETs, this will provide Jim with options when the teeth fitted need replacing.

HSM has years of experience in making rippers and has matched the ripper model size to the performance

One of the reasons for selecting the R960SME is down to Jim Jamieson's good experience with a pre-owned R954 machine.

The R960SME has a massive undercarriage and can handle operating over the tracks or over the side.

HSM buckets are made from Hardox Swedish steel and chromium carbide is used throughout to give it extra strength and reduced wear.

Liebherr quick hitch is made in the mining division next door to the main factory in Colmar, France.

The view through the rock guard is excellent.

of the R960SME. The ripper has a considerable number of wear plates fitted and is made from the same 450 Hardox steel. It uses the ESCO SV2 tooth retention system, which is claimed to have longer wear life and greater productivity from a slim nose profile, with up to 72% useable wear metal and up to 30% increase in nose strength.

As an aside, HSM has started fitting its logo, in the shape of an ant, on to the side of their attachments, to represent lightness and strength. Apparently an ant can carry 50 times its own body weight – and the 1,800mm wide, 3.3 cu.m capacity heavy duty bucket weighs in at a respectable 3,750kg.

At the heart of Jamieson's R960SME beats a powerful 11.9-litre six-cylinder turbocharged Stage IIIB/Tier 4i diesel engine, which produces 350hp at 1800rpm. With no AdBlue after-treatment, only a diesel particulate filter is fitted to achieve these emission standards, as

this particular R960SME was one of the last machines to be built with the Stage IIIB engine, as Liebherr has now exhausted its stock. Going forward, the R960SME will have the Stage IV power unit that will be fitted with SCR and urea injection (AdBlue), which will still produce the same power output.

The R960 has a new slightly roomier cab and has a host of standard equipment to provide the driver with a comfortable place to spend a shift, as the R960 has low interior noise levels at a claimed 72dB(A), automatic air conditioning and a weight sensing pneumatic seat with heating fitted as standard. The driver also has an excellent view to the front, despite the rock guard fitted, and the view to the rear is good as a result of a low-profile engine cover and a rear-view camera linked to a 7in high resolution colour touchscreen monitor.

Liebherr has fitted two robust steps on the side of the undercarriage to access the cab and upper structure.

This allows good access to reach the top of the upper structure via a door panel and inboard step – between the cab and power pack – (that also acts as a storage box/toolbox) to allow the driver to manually check the engine oil and coolant levels. However, if you are prepared to trust modern electronics, these levels can be checked on the LCD screen mounted in the cab. Refuelling the large capacity 770-litre fuel tank is via an electric fuel pump, which is located in the offside storage compartment, where the auto-lube reservoir is also located.

With the driver walk-round checks completed, the R960SME was tracked from the workshop area on to a pad of blasted rock to begin digging the fragmented material and swinging right through 90 degrees to create a stockpile of material for a Cat 336E excavator to load the primary crusher, a Terex 1480 mobile jaw crusher. It was feeding a secondary Terex 1550 mobile cone crusher, with the material finally reaching a five-way Terex 694 mobile screening unit.

At the time of our visit, the first operator to get behind the controls of the R960SME was Graham Webb, who was currently driving a 38-tonne Volvo excavator. That said, he had also put a significant amount of hours on the R954 machine, so he knew what a big excavator feels like to operate. He was happy with the R960SME's capabilities and he thought there was a marked improvement in performance compared to the R954. He felt the R960 is more sure-footed and has good digging performance due to the higher breakout forces that an SME spec machine produces, and despite its size it was performing three passes in under a minute.

Working with the screener units are three CAT wheeled loaders, the newest and biggest being a 287hp, 25-ton Caterpillar 972H; this front line machine is fitted with a 4.3 cu.m bucket to carry out stockpiling and HGV truck loading duties. The 972H has also been fitted with a bespoke bucket with modified side plates, teeth segments and wear plates to increase the durability and back-guard to slightly increase productivity, as I witnessed the 32-tonne eight-wheel tippers being filled in three passes. Plus, the standard tyres have been swapped for a more durable set of Michelin XLDD 2A boots to deal with abrasive rocks found in this application.

Ardlethen Quarry produces on average 500,000

R960SME has been nicknamed the nut cracker, because of its rock ripping duties.

Jamieson operates a large fleet of Cat wheel loaders and excavators.

tonnes of crushed rock each year, with the main output used by the major house building companies, and with construction work on the A9 Aberdeen bypass project, the quarry should see an increase in work, as it has already had a special order for some 250mm and 40mm stone that was thoroughly batch tested prior to delivery on site.

Brian Clark is the transport manager, based at Ardlethen, and is responsible for Jim's stunning 23-strong fleet of modern Scania 32-tonne eight-wheel tippers, and a 730hp V8-powered Scania tractor unit, plated at 150 tonnes, which was coupled to a four-axle steering Nooteboom Euro 83-04 low-loader to move the R960SME and the mobile crushing/screening plant. With the HGV fleet

The large 110-acre site has sufficient reserves of stone for the next 14 years.

based at the quarry, Jim's team is able to provide a one stop shop, and with the new R960SME now operational, it can react even quicker to meet customer demand, which can require it to fill a 6,000-tonne order of stone within a week, at extremely short notice!

Jim takes a great deal of pride in all of his trucks and plant equipment, and with that in mind, the R960SME has been fitted with a comprehensive set of very eye-catching decals, printed and cut from the latest 3M reflective tape. Along with his mobile crushing and screening business, Jim's customers will no doubt get to see this latest investment in his modern fleet of equipment in due course.

CHAPTER 25

CA Blackwell (Contracts) • Cat 992K • Yeoman Glensanda super quarry • September 2015

The 2,400ha Glensanda Estate is located on the shores of Loch Linnhe in the Western Highlands of Scotland. Some 420 million years ago, with the formation of the Caledonian mountain range, molten rock cooled to form extremely hard granite.

In 1982 Foster Yeoman began work to create a coastal super quarry at Glensanda. Without the benefit of road access to the site, everything had to be shipped across the sea loch. Operations commenced in 1986 and the site has since been acquired by Aggregate Industries.

Yeoman Glensanda is Europe's largest granite quarry and, together with its massive Norwegian coastal quarries at Halsvik and Kvalsund, provides Aggregate Industries with a truly massive reserve and production capacity. Glensanda alone has estimated reserves of 760 million tonnes, which at current production rates will last for the next 100 years. Glensanda supplies high quality crushed granite to markets across Europe and occasionally to North America. The output from this site accounts for close to 100% of all aggregates exported from Britain and Glensanda is one of the UK's largest ports by tonnage. It is estimated that it generates £100 million per year for the Scottish economy.

Staff are shuttled back and forth on a private passenger vessel from the port of Barcaldine, 6 miles north of Oban. Adjacent to the passenger terminal there is a large jetty to accommodate the vehicle transporter, which is capable of carrying a pair of 100-tonne capacity haul trucks, but is usually used to transport the host of commercial vehicles required to service the site.

The 45-minute trip across Loch Linnhe on the passenger vessel passes the Isle of Lismore before landing at Glensanda's harbour. The quarry dock has sufficient draft – in a 200m-deep sheltered channel – to accommodate 100,000-tonne ocean-going bulk carriers. At the time of my visit 85,000 tonnes was being loaded on to the MV Yeoman Bontrup, which stands at a massive 250m long and 38m wide.

When this vessel reaches its destination – in this case the London Gateway development – it can self-discharge its cargo at a rate of 6,000 tonnes per hour. This is quite an achievement for Aggregate Industries, delivering the equivalent of 4,500 HGV truckloads of high quality granite to within 20 miles of central London by sea direct from the quarry.

As all the granite leaves the quarry by sea, it is

Glensanda super quarry, with Oban in the distance, 11 miles away as the crow flies.

One of the top benches at 520m above sea level. Note the short haul to the primary crusher shed at the top. Loch Linnhe, Lismore and Oban can be seen in the backdrop.

A 97,000-tonne cargo ship *Yeoman Bontrup*, owned by Aggregate Industries, bound for the London Gateway port.

important to note that it is not all about huge shipments. Local companies can charter cargo boats to pick up part loads and deliver the material to the Scottish islands and other ports around the UK.

In July 2013 CA Blackwell (Contracts) was awarded a four-year load and haul contract for the site, shifting approximately seven million tonnes of blasted rock each year from the quarry face to the processing plant. Our guide for the day was Craig Smith, senior production manager for Blackwell at the site.

Drilling and blasting operations are subcontracted to BAM Ritchies. Although the benches are nominally 20m deep, the drill rigs bore 22m holes. The first extra metre covers the 15-degree angle of the face and the other extra metre ensures the material is fragmented below bench height, to avoid damaging the wear parts on the loader's bucket. Explosives are supplied by Orica – about 1,500 tonnes of emulsion being required each year on this site – and transported to the blast area by a Cat 25-tonne articulated hauler chassis fitted with a Tradestar tanker body and pumping unit.

Before I can head up the mountainside to see the big Cat prime movers in action the three-minute blast warning is sounded. Then a modest blast sees around 20,000 tonnes of high quality granite land on the quarry floor. With the blast over and the all-clear given, we head up the smooth haul roads to find the first big Cat 992K wheel loader.

The site is a long-standing user of Caterpillar equipment, as is Blackwell, so Finning was the obvious choice to supply the machinery for the load and haul contract. A pair of 100-tonne 992K wheel loaders and a 6015 backhoe excavator have been supplied, together with eight 777G 100-tonne capacity dump trucks. As tried and trusted prime movers on this site, it is worth noting that Cat 992s have been in production for nearly 50 years. Glensanda used a C Series model from the start of operations, then moved on to the 992D in 1993, then 992G machines were shipped to the quarry in December 2007. The 992K loaders replaced the G Series models with 20,000 hours on the clock during November 2013.

CA Blackwell (Contracts) has kitted out the 992K loaders with extra metal to withstand the rigours of

Orica supplies the explosives, which are transported to the drill site using a converted CAT articulated dump truck.

Haul road dust is kept to a minimum with a 9.000-litre CAT dust suppression unit.

loading the highly abrasive granite. The 10.7 cu.m heavy duty rock bucket has more than 3 tonnes of extra wear protection fitted. However, due to the nature of the material, a significant volume of this armoured steel will be worn away and the buckets require regular maintenance. The bucket also has an extended spill board to provide extra protection to the hydraulic and electrical components at the back of the bucket and the linkage. The high spill board has slots cut into it to help the operator achieve a good fill factor.

The Cat 992K wheel loaders are fitted with giant 45/65-R45 Michelin XLDD2 tyres, which are protected by 22mm-thick RUD steel chains on both the front and rear. Tyre maintenance is one of the most expensive operating costs on such large loaders in this material, so fitting such chains is critical to extending their lifespan. Incidentally, without the chains fitted, a single rim and tyre of the 992K wheel tips the scales at more than 3½ tonnes.

Blackwell's operators will spend 10½ hours in the cab of the 992Ks, with two half-hour rest breaks. They have a target time of 2½ minutes using four or five passes to load the 777G haul trucks with 100 short tons of high quality granite. The big Cat shovels have sufficient bucket pin height, at just under 7m and reach, to achieve a good heaped load in the skips.

Without a steering wheel on the 992Ks, the view to the front is excellent, as the joystick controls both the steering and gear shifting. The force feedback steering controls and hydraulic levers are integral parts of the air suspension seat, which has six-way adjustment to accommodate all sizes of operator.

At the heart of this massive 15.7m-long wheel loader is Cat's tried and trusted Stage II emissions compliant V12, 32-litre engine, which in this application produces 904hp at 1750rpm. The 992K has the same cast box boom design first seen on the previous G Series machine. Cat says that this delivers three times the torsional loading strength of the previous steel plate lift arms. The narrow mono boom is fitted with maintenance-free sleeve

Benches are nominally 20m deep, the drill rigs bore 22m holes. The first extra metre covers the 15-degree angle of the face and the other extra metre ensures the material is fragmented below bench height, to avoid damaging the wear parts on the loader's bucket.

bearing pins to help reduce maintenance costs and a single tilt cylinder, which also helps in providing the operator with a good view of the bucket.

At the time of my visit one of the 992K shovels was working an old face some 500m above sea level where some of the rock was weathered and cracked. When blasted, such material is easily dislodged, resulting in it not being fully fragmented. The 992K was moving the oversized stone to one side to allow a nearby excavator fitted with a hammer to reduce the lumps to about a metre square, to match the in-feed capacity of the primary crusher.

Craig was acting as our tour guide, but he was also keeping a close watch on production levels. This part of

the site was starting to run out of blasted stone, so he decided that it was time to move the 992K to a freshly blasted area of the quarry.

Craig explained the 100-tonne Cat wheel loaders can be quickly deployed to other parts of the quarry and are capable of shifting a massive amount of material. While it is not uncommon for the blast team to dislodge more than 80,000 tonnes of granite on to the quarry bench in one hit, the 992K will shift this amount in just eight days! To put this into perspective, some hard stone quarries are proud that they can crush up to 6,000 tonnes in a normal working week. The production team at Glensanda will achieve this level of output in two to three hours.

The other 992K was working at the opposite end of this vast quarry in a section known as 420 bench (the height in metres above sea level). Here the 777G trucks were hauling stone to the primary crusher on the longest haul of the site, covering 4.4km in one return cycle.

This part of the quarry was being expanded with the help of Blackwell's 50-tonne Cat D9T dozer, which was ripping and dozing the bench to level in preparation for the next blast. This was to help ensure there is no damage to the 992K's bucket teeth when it enters the blasted material. Despite the hard granite material, it was an incredible sight to see the D9T's ripper tooth slice through the ground with little or no track slip.

The 91-tonne capacity Cat 777G haul trucks have also been modified to deal with the hard granite material, with Hardox liners fitted to the load area of the skips. In order to improve load retention, the sides of the skips have been fitted with spill boards, while angle iron bars fitted on the skip's canopy help to prevent material sliding across the flat canopy roof when travelling on the haul roads.

The new 777Gs are powered by the same C32 engine found in the 992K, only in this application it is rated at 1,025hp at 1800rpm. In addition, in the 777G model, the engine is certified to the equivalent of Stage IV emissions compliance, thanks to a pair of diesel oxidation catalysts that feature passive regeneration, with no need for diesel particulate filters nor AdBlue. Cat claims that the G Series 777s have a significant number of improvements over the F series trucks. The big V12 engine produces 7% more torque and is mated

Cat D9T is preparing and ripping to allow the Cat 992K to work this area without damaging the bucket teeth.

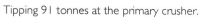

Tipping 91 tonnes at the primary crusher.

to a new transmission control that uses part throttle shifting and torque shift management. This is said to deliver quick haul cycle times with automotive-quality shifting and comfort. The traction control system has also been improved as it is now steering sensitive to differentiate between tyre spin and high-speed turns. It also provides improved performance when pulling off the bench in slippery conditions.

Improvements in the design of the cab have reduced the sound levels by a claimed 50% for the operator. The 777G also has a tyre monitoring system fitted as standard, which measures the ambient air temperature, the truck's speed and distance travelled, and the manufacturer's rating for the tyres. If the tyre approaches its calculated limits, the operator receives a warning inside the cab. To help the 992K operators achieve the target payload of 91 tonnes for the 777G trucks on this site, each hauler is fitted with a large exterior LED screen that displays the payload after each pass.

The load and haul operation is closely monitored by Blackwell with the help of Cat's Minestar telematics system. This produces a large range of production reports, covering target output vs actual, as well as target times and individual shovel and truck performance. This data is available not only to management, but is also displayed in the quarry canteen for all to see.

During my visit I was invited into the cab of a 777G operated by Jamie Talbot. Jamie had been a chef for 27 years before joining the Glensanda team eight years ago and obviously enjoys driving 165-tonne gross weight trucks on a four days on, four days off shift pattern.

Jamie still has fond memories of the 777D model he started out on, but is full of praise for the latest G Series model and said: "The 777G is a very good truck, really comfortable from a driver's point of view and performance wise; they certainly pull the stone up the hill and that's what they're here to do."

The cab is very spacious and comes complete with a very comfortable training seat. The main difference between the cab of the old D Series and 777F/G models is that the operator's position has been moved inboard to a more central position. This moves the operator away from the direct shocks of the nearside front wheel station. However, if Jamie was being picky about the

The two Cat 922K wheel loaders and the eight 777G trucks were prepared in Glasgow, Finning's HQ in Scotland. Note, the cab is removed to transport it under bridges.

The 992K bucket pin height has plenty of dump/pin height to achieve a heaped load in the 777G trucks. Due to the abrasive granite, 3 tonnes of extra wear protection plating is added to the big 10.7 cu.m bucket. Note the rock deflectors on the bucket side plates.

The big wheel loaders have high mobility and are well suited to a site this size. Photo: Peter Haddock, of Edson Evers PR

Cat 777G haul truck heading to the crusher with a full load of granite.

View from 777G to the new workshop as it heads back for another load.

777G, he would like to see a wiper on the cab door window to keep it clean during inclement Scottish weather.

Once loaded, we head up to the primary crusher and enjoy the 777G's low noise levels. On arrival at the crusher at more than 500m above sea level we have magnificent views over the west coast of Scotland. When it is our turn to tip, the 777G's skip is raised in 11 seconds and lowers in 7 seconds, before heading back down the haul road for another load. Any oversize rock spotted in the crusher is dealt with by a remote-controlled Cat excavator fitted with a ripper tooth to the dipper arm.

Paul Bird, project manager and Blackwell's project manager on site said: "The Cat 992Ks and 777G trucks are working well, exactly to expectations. We have set targets with Finning for availability and those are being met or exceeded, and the newer models are more efficient, which has helped to reduce fuel burn."

Blackwell also operates the only Cat 6015 in the UK. This 110-tonne backhoe excavator is used for digging out tight pockets of stone and to release any loose rock that did not drop during blasting and would therefore pose a safety risk.

The site is too large for the 6015 to track to each job, so Blackwell has invested in a Sleipner transport system from Finning to overcome this issue. It comprises a set of ultra-heavy duty wheels placed under each track of the big backhoe, which is then pulled by a 777 truck as the bucket of the 6015 rests in the part-loaded skip.

At the time of my visit the 6015 was in Glensanda's brand new four-bay workshop for maintenance to the stick and engine module. In charge of maintenance is Blackwell's Martin MacDonald who, together with his team of 13 people, is responsible for maintaining all the mobile plant on a 24/7 basis. Finning also has a full-time maintenance technician on site to look after the Cat equipment that is subject to a full repair and maintenance package. The only items not covered under this service contract are tyres, tyre chains and ground engagement wear items.

Martin explained he always keeps a good stock of bucket wear items, as the 992K shovels, teeth, corner-shrouds and blocks wear quickly, requiring regular replacement.

The bigger four-bay workshop – staffed by Blackwell's own engineers and one fitter from Finning – has been future-proofed to take bigger loaders and trucks should production need to be increased. The Cat 6015 was in the shop for some maintenance work.

However, much more frequent work is required on the 6015s. Martin is pleased to report the 992K and 777G models are running at 98 to 99% availability and says: "The 777G is a good truck and a big improvement over the previous model, and the 992K is just outstanding, everyone has a great deal of respect for it."

Blackwell also operates a Cat 16H motor grader and a 972K wheel loader to help maintain the haul roads. Dust suppression is taken care of by Blackwell's converted Cat articulated dump truck now hauling and spraying water. In addition, a Cat telehandler is used to support the workshop activities.

The new workshop, just six weeks old at the time of our visit, has been designed to be future-proof as it can accommodate larger machines, for example Cat 994 loaders and Cat 789 trucks, as a potential load and haul combination for the future.

Production capacity at the Glensanda super quarry has been increased with the addition of a new primary crusher, which will up crushing rates from 2,250 tonnes per hour to just over 3,000. In addition to the crusher, Aggregate Industries' recent investment of £30 million includes a new reclaim tunnel. The new tunnel will connect to the new 230m drop Glory Hole with its 1.8km tunnel leading to the 500,000-tonne surge pile and the secondary processing equipment.

As well as increasing production capacity, the new crusher is located 60m below the existing unit, which stands some 520m above sea level. Not only will the location of the new crusher result in shorter haul times but the removal of the original unit and its conveyor from the skyline will also enhance the visual appeal of the quarry. In addition, the space occupied by the old crusher can now be mined for millions of tonnes of high quality granite.

With a new workshop, a new primary crusher and Blackwell's large wheel loaders and haulers delivering fantastic availability, Aggregate Industries is able to meet market demand for high quality granite well into the future.

Roads

In this section I cover two major roads infrastructure projects and one road widening and rail bridge construction project.

The Forth Replacement Crossing is a massive project in which the Scottish government, through Transport Scotland, is spending £1.45 million to build an additional road crossing over the Firth of Forth at Queensferry, as the existing road bridge main cables have lost some of their strength and the bridge closes to high-sided vehicles (double decks buses) once wind speed exceeds 45mph and to HGVs, bikes and pedestrians when gusting above 50mph.

The other is the £500 million M8 M73 M74 Motorway Improvements Project. This will include the construction of 12km of new M8 motorway and a new A8 all-purpose road, incorporating upgraded sections of the existing A8, which will provide uninterrupted motorway and cut 18 minutes off peak journey times between Glasgow and Edinburgh. The project also includes a major redesign of Raith Junction, one of the busiest strategic connecting junctions within central Scotland's motorway network.

Both of the above projects were covered by spending just one day on site, however, I spent three months – with an hour here and there – following the tunnelling through a railway embankment, extending a bridge on the A723 between Holytown and Carfin to eventually create a dual carriageway link from Motherwell to the M8 motorway at Newhouse. This individual project is small in comparison to the others covered in this chapter, at £10 million, but is part of a 30-year, £1.2 billion, redevelopment of the former Ravenscraig Steel works that closed in 1992. This project was interesting, as the 1,800-tonne bridge being constructed would need to be lifted and carried into position by special trailers in order to minimise disruption to the rail passengers and goods trains from Glasgow to Edinburgh.

A number of earthmoving machines on these sites are using some of the latest GPS machine guidance equipment to save time and money, and improve safety for ground workers. The health and safety (H&S) standards on these sites are some of the highest I've come across, particularly when dealing with railway work. In fact, it is fair to say that H&S culture across all four applications covered in this book is to a very high standard.

CHAPTER 26

Forth Replacement Crossing • Firth of Forth • July 2014

As far back as 1130 pilgrims travelled north regularly across the river Forth to St Andrews to worship at the site of the saint's relics by boat, and for more than 760 years, ferry boat was the only form of transport used to cross the Forth.

The first bridge to span the Forth, at Queensferry, was the iconic Forth Rail Bridge, which cost £3.2 million, used 65,000 tonnes of steel, took 4,600 people seven years to build and hammered home 6.5 million steel rivets. At 1.5 miles it was the longest cantilever bridge in the world until 1917. The bridge was opened on 4 March 1890 by the then Prince of Wales, who drove home the last rivet, which was gold plated. The bridge

provided a continuous East Coast railway route from London to Aberdeen.

The first road bridge was opened on 4 September 1964 and it took six years to construct, at a cost of £19.5 million. It replaced the cart and later car ferries that had serviced this historic route for hundreds of years. At 1.5 miles long, it was the UK's first long-span suspension bridge, and it was also the longest suspension bridge outside the USA when it opened. After more than 40 years of high volume traffic crossing the structure, investigations showed the suspension cables to have lost 10% of their original strength. There were also ongoing issues of high winds, which close the bridge to high-sided

Standing at the top of the highest motorway embankment looking north to the Ferrytoll junction with the viaduct columns in the foreground.

vehicles, and it has no hard shoulder to accommodate vehicle breakdowns for the 24 million vehicles crossing per annum. Taking all this into consideration, it was formally decided in March 2011, with the passing of the Forth Crossing Act, that a new road crossing would be constructed.

The Forth Replacement Crossing is a cable-stayed bridge that is being constructed to the west of the existing Forth Road Bridge. It is the largest civil engineering project in Scotland for a generation and will replace the Forth Road Bridge as the main vehicle crossing on the Forth.

The ability to complete such a large project normally requires a consortium of companies. Forth Crossing Bridge Constructors (FCBC), an international consortium consisting of Hochtief Solutions, American Bridge, Dragados & Morrison Construction, won the tender to build the new crossing. Work began in the summer of 2011, and at the time of my visit substantial progress had been made. The estimated cost of the project had been reduced by £145 million to £790 million for the bridge section and the project is part of a package of transport improvements costing £1.45 billion in total. Overall, this money includes upgrading the infrastructure both north and south of the new crossing.

Scotland's then First Minister, Alex Salmond, was on site to announce the name of the new bridge – which was chosen by members of the public as the Queensferry Crossing.

My tour of the site started from FCBC's project office block on the north side, close to Rosyth docks, and a short distance away I found a new 31-tonne, 213hp, Doosan DX300 LC excavator working on a section of the upgraded Ferrytoll roundabout. At this particular site, drilling and blasting had taken place to move 120,000 cu.m of fill for a 25m high road embankment – on both sides of the crossing – in a massive cut and fill operation whereby all the excavated material is being reused, particularly in areas that required new road sections constructed on marshland.

At the Ferrytoll junction, concrete columns will hold a new viaduct section of motorway, where the construction crew has cast a slab on to the existing rock bed and used reinforced steel starter-bars to build 14 1m diameter columns to support a capping beam for steel box section beams to sit on. Once the beams are

in place, a concrete road deck section will be laid on top of them. Nearby, similar work is taking place to make the highest motorway embankment – at 25m tall – in the UK, and on this section of the site a Casagrande C6 drilling rig is taking test bores to ensure the concrete piles do not contain any air pockets. If found, the machine will be used to pump concrete to fill and repair any voids.

Over on the south side of the Forth is the formation of the new M90 extension, with the road standing more than 5m high from the original land surface. Material for this has been won – as part of the cut and fill plan – from both north and south construction areas. Near to where the new south viaducts are being constructed, a 30-tonne JCB 290LC 360 excavator was quickly loading numerous eight-wheeled tippers with shale and then transporting it a short distance to form part of the base of the M90 extension. There, a 150hp, 36-tonne, Cat D6N dozer was spreading the material, and a Bomag BW213 BH, fitted with a sheep's foot roller was compacting the surface. An 18-tonne Cat D4H dozer carried out final trimming and another Bomag machine, this time fitted with a conventional smooth drum, was compacting type one material prior to the black-top being laid. This section of the road also has a 3m high embankment to act as a noise shield for the adjacent houses.

At the end of the M90 extension, at the north approach viaduct launch pit, is a massive 120-tonne capacity Van Haasan overhead gantry crane, which is 35m wide by 20m high. It is being used to unload steel viaduct sections from HGVs arriving from Cleveland Bridge UK Ltd, a Darlington-based engineering company. These sections consist of 72-tonne steel viaducts that come in 33m sections and are pre-painted with a zinc primer and four coats of grey epoxy resin paint. Once on site, the sections are welded and bolted together into longer sections up to 90m in length before final preparation for launching out over the V-shaped support piers using hydraulic jacks.

On the southern approach viaduct, powerful hydraulic machines, called strand jacks, are fixed to the bridge abutment and connected to the viaduct sections via steel cables. This jack and cable system then slowly pulls the sections out towards and over each support pier, which will reach a combined length of 543m (on the south side) and 202m to cover the north span. Further strand jacks are attached to vertical 'king posts' to lift

One of the 110 road deck sections being lifted into position with a strand jack system (taken in January 2016). Photo: William Spence

the leading edges of the structures to counteract the effect of gravity and maintain the correct height to slide over the top of each pier. FCBC believes this is the first time king posts have been used in this way in the UK. Once fully assembled they will have total weight of 5,000 tonnes, and carry all north and southbound traffic to and from the bridge. They are a vital part of the new cable-stayed main crossing, connecting the new bridge to the land on either side.

The main bridge itself, with its long spans, massive foundations and three huge concrete towers, performs the function of crossing the deeper water in mid-estuary while leaving navigational channels for shipping open.

Back at Rosyth docks an 18-tonne, 217hp, Caterpillar 950H wheel loader was busy keeping the specially designed concrete batching plant hoppers full. This plant is fully computerised and automated, with FCBC claiming it is one of the most modern in the UK. The concrete is batched here before being shipped out on barges to the middle of the Forth in a continuous operation that can

involve up to 100 people. Much of the raw materials that make up the concrete – the sand, aggregates and water, for example – are being sourced locally. When working flat out the plant will be producing up to 120 cu.m of concrete per hour, which means a HGV truck loaded with 8 cu.m will be leaving the plant every four minutes! The barges were specially fitted out for FCBC by a local company, Briggs Marine in Burntisland, with six concrete mixers to keep a total of 72 cu.m of concrete mix 'live' as they are pulled by tugs out to site.

Building the three towers involved the creation of huge concrete plugs inside the foundation caissons and a central tower cofferdam. This operation was launched in late July 2013 with 7,400 cu.m being successfully poured into the north tower caisson. This was followed, in early August, by a 4,400 cu.m concrete pour into the centre tower cofferdam.

Finally, in late August 2013, the team made a world record for the largest continuous underwater concrete pour of 16,869 cu.m in just 364 hours; this went into the

A view taken in July 2014 from the south shore.

An image taken 18 months later (in January 2016) from the south shore with some road deck sections in place on the three towers.

After the world record breaking concrete pour, the focus now shifts from below the waves to the upward construction of three towers above the waters of the Forth (taken during November 2013).

Powerful hydraulic machines, called strand jacks, are fixed to the bridge abutment and connected to the viaduct sections via steel cables. This jack and cable system then slowly pulls the sections out towards and over each support pier.

A JCB 290LC excavator loading eight-wheeled tippers with shale and then transporting it a short distance to form part of the base of the M90 extension.

Cat wheel loaders work hard to keep the batching plant operational day and night.

A barge heading back to the quayside for another 72 cu.m load of concrete.

south tower caisson to form a solid plug more than 26m in depth. The average pour rate for the south tower caisson was 47 cu.m per hour, with the total weight of the concrete almost 39,000 tonnes, the equivalent of 3,250 London buses, and the four barges covered a total distance of 1,800km, the approximate distance from Land's End to John o'Groats!

At the time of the pour, Carlo Germani, FCBC project director, said: "The underwater concrete pour operation has gone without a hitch thanks to extremely detailed advance preparation carried out by the team. This achievement is a credit to the skills of everyone involved. It is a huge milestone for the project because it means that the focus can now switch from below the waves to the upward construction of the three towers above the waters of the Forth."

During April 2013 the first of three identical 40-tonne capacity Liebherr 630ECH 40 cranes were supplied and erected by Streif Baulogistik, one of Europe's largest tower crane rental companies, and these will help to build the three main pylons to a height of 207m. As the pylons are built, the crane will self-climb alongside them, by lifting its own vertical tower sections and using an external climbing unit to support the upper structure/ slewing frame. The company is planning to climb the cranes in eight phases, reaching a final hook height of 226m, or 741ft in old money!

At the time of my visit, the cranes had helped build the towers to another milestone, in reaching the 60m deck height level. Preparations were being made to install the first steel box sections that will eventually support the road surface and as the towers continue to rise towards full height, additional sections will be lifted into place from barges using mobile derricks located on the bridge.

The fabrication and delivery of all 110 sections of the deck steelwork was awarded to Shanghai Zhenhua Port Machinery Company (ZPMC) of Shanghai. The standard pre-fabricated sections are 16.2m long by 40m wide and each of the steel sections weigh between 200–350 tonnes before reinforced concrete decks are cast – using

The 1,200-tonne lift capacity TAKLIFT 6 is placing a 120-tonne trestle into place in preparation for the first 750-tonne road deck sections (September 2014).

concrete produced in FCBC's batching plant next door – on top. This will increase the weight of the segments to a massive 750 tonnes. The last part of the yard operations is to roll them out, using a self-propelled modular transporter (SPMT) system, and on to a transport barge, which will take the deck sections out to the towers at a rate of two segments per week.

The largest excavator on this project is one supplied by WM Plant hire; a Caterpillar 375D long-reach machine, weighing 115 tonnes and with a maximum reach of 32m. It was hired to excavate all the sediment (mud, alluvium and glacial till) found at the bottom of the caisson. The 375D was working off a barge, at S4 support pier location, just off the south shore line, and

can only achieve its full reach when working at high tide, otherwise the excavator's boom foot will hit the top of caisson steel piles.

The team behind the new crossing has gone on record to provide an assurance that, unlike its sister the current Forth Road Bridge, it would remain open no matter the weather, as the new bridge will be fitted with special barriers that prevent it from having to close in a storm. David Climie vowed: "The windshields have undergone wind tunnel testing and there will be no need to close the bridge in high winds."

This project is a truly international operation, but that said, it is also supporting the local economy with around 250 out of 415 subcontracts having gone to Scottish companies and 75% of the 1,000-strong workforce from north of the border. Local firms have been awarded subcontracts or supply orders with a total value of about £157 million.

Many thanks to FCBC's communication team for supplying technical information and aerial photography showing the M90 extension, new Queensferry junction and existing A904 (rerouted).

Since my official visit, I have continued to follow the crossing's progress, and have included some images with some road deck sections in place during January 2016.

August 2014 – The M90 extension, new Queensferry junction and existing A904 road (rerouted). Photo: Transport Scotland

CHAPTER 27

Road widening on A723 and Network Rail project • Motherwell • July 2014

This project is part of a 30-year, £1.2 billion redevelopment of the former Ravenscraig Steel works that closed in 1992. This latest £10 million of funding, by North Lanarkshire council, is paying for tunnelling through a railway embankment, extending a bridge on the A723 between Holytown and Carfin and securing small parcels of land along the A723 required for road widening. The aim is to eventually create a dual carriageway link from Motherwell to the M8 motorway at Newhouse.

BAM Nuttall is the main contractor for this project and after creating a works compound consisting of pristine offices, canteen and car park, the first phase of the project was to remove some 45,000 tonnes of material, a mixture of soils, rock and parts of a 0.5m thick coal seam, which is situated just 10m below the natural topography! This site was to form a wide base for bridge construction, which was to be built adjacent to the railway, and then moved into position, to keep rail passenger and goods traffic disruption to a minimum. The material was excavated by three 20-tonne class 360 excavators, a JCB JS200 and a Komatsu 210. Blasting

During June 2014 a total of 45,000 tonnes of muck and rock was being shifted.

of rock was ruled out, as the site is surrounded by residential housing and BAM is part of the considerate constructor scheme and keen to keep noise to a minimum. Instead, a Hyundai 20-tonne machine fitted with a ripper tooth was used to prise out the large slabs of stone, which would not look out of place sitting in the bucket of a 300-tonne class monster miner!

This first phase of muck-shift produced approximately 16,000 tonnes of soil and 3,000 tonnes of rock to be left on site for reinstatement and help provide the final profile of the large carriageway batters. The remaining 26,000 tonnes of excavated material was to be transported to Glasgow, using a fleet of HGV tippers, to be used on another construction project.

Three weeks into the project (mid-June 2014), and with the site cleared and level, the reinforcing steel bars were formed and a team of joiners set about building the wooden shuttering to form footings for the new bridge. The bridge was to be moved into its final resting place by a self-propelled modular transporter (SPMT) system, and with that in mind the bridge footings were designed and constructed using numerous shearlinks – 12mm diameter reinforcement bars – that were used to tie all the main steel bars (which range from 32, 25, 20mm in diameter) together to make the structure much more stiff for the lift and carry operation.

Site manager Frank Young explained: "If we were building the bridge in situ, then the specification would have required a lot fewer reinforcement bars; in the footings and where the abutment walls meets the roof these have a significant amount of extra reinforcement to deal with the strain during the lift and carry operation."

Once the shuttering was completed, 14 truckloads of C60-50 (60nM strength after 50 days) hi-flow spec concrete was poured using a 24m reach Schwing pumping unit, which can pump 70 cu.m per hour. Each 8 cu.m truck load of concrete was batch tested on site and had to meet a slump rating of between 80 and 180mm before it was allowed to be unloaded into the hopper of the concrete pumping truck. Test cubes were also made and were tested to destruction to judge the rate at which the concrete set before the job could progress on to the next stage of construction.

Towards the end of July and into the first week of August, the sides and roof reinforcement bars of the bridge were craned into position along with the subsequent shuttering. Fourteen extra layers of re-bar were also used at the roof and main wall abutment to cope with the lift and carry operation.

Simultaneously, on the other side of the railway line, preparations were being made to grout an old (capped) mineshaft situated close to the final resting place of the new bridge. BAM subcontracted the grouting to Van Elle and it was using one of its 11.7-tonne Casagrande C6 drilling rigs to assess the ground conditions of the old shaft and provide grouting boreholes. Since the drilling rig unit, with its 9,922mm high mast, was working so close to the railway, BAM took possession of the track – on a night shift basis – whereby all train movements were stopped. This was a precautionary move, just in case there was unforeseen subsidence of the mineshaft. That said, BAM took no chances and deployed a Bailey bridge platform to straddle the void to ensure the machine remained stable throughout the drilling operations.

At the time of my visit, Lydia Stevenson, assistant engineer, was helping to set out the final drilling positions of the mineshaft. Lydia is a civil engineer graduate of Strathclyde University and had spent two years on summer placements with BAM Nuttall as part of her course work, leading to a full-time job.

During the nightshift the drilling operation went to plan, with the first 40m performed by simply pushing each 3m long rod through the mineshaft void, and for the remaining 80m the hammer head on the machine was required. The last 3m of the 119m of drilling was hard going, confirming the team had reached the bedrock (not just single slab of rock). Over the next few weekends, four more holes were drilled, followed by casings inserted in order to pump 160 tonnes of 10:1 mixed fluid PFA and Portland cement grout through the entire length of the mineshaft. After each phase of drilling operation, a team of engineers from Network Rail checked the adjacent track gauge dimensions for any possible movement – nothing was taken for granted before the line was reopened at 7.30am.

One could not help but notice a strong safety culture and high level of detail to keep the job and everyone safe, from pre-shift briefings, full HiVIS PPE, to portable steps with handrails to cover off slight inclines right across the site, and other safety initiatives far too numerous to list.

During the last week of August, the four wing-walls were being formed and a 31m reach Schwing concrete

A FLASS (Fine Lining and Sleeper Spacing) machine from McCulloch Rail is deployed to lift the heavy concrete sleepers.

pumping unit poured 23 cu.m of concrete into each of the wings – the same high strength concrete material as used in the main walls and roof. And by the first week of September all the wing walls concrete had set and shuttering scaffolding stripped away.

BAM subcontracted all rail track activity to Babcock Rail and at the start of September it had piled and fitted a temporary 52m long span frame to support all the control and communication cables that lay adjacent to the track. Babcock had also fitted a new high voltage pylon further up the line to allow the removal of the existing pylon, which was situated where the new bridge was due to rest.

On Saturday, 20 September, BAM took possession of the railway line for the next nine days and the job reached a very time critical phase as the clock was now ticking to reopen the Glasgow to Edinburgh line for passenger and goods trains. At around 1.30am machines moved in to cut and lift a section of steel rails, and then a FLASS (Fine Lining and Sleeper Spacing) machine from

McCulloch Rail was deployed to lift the heavy concrete sleepers from the area of the cut. During the nightshift, two 30-tonne excavators were deployed, one on either side of the railway embankment, and they shifted around 20,000 tonnes of material over the next two days. On the south side was a Doosan DX300LC swinging a 1.3 cu.m rock bucket to load a steady stream of 25-tonne capacity Cat 725 articulated dump trucks (ADTs). The ADTs ran the material up a one-way system to the top of the site and this material was stockpiled and reused as part of the reinstatement phase. At first light (7am) with all the railway workers now off the track embankment, the team could start to speed up the dig and deploy a 28-tonne Hitachi Zaxis 280 LC excavator to help load the ADTs.

Over on the north side of the line, a new 30-tonne JCB JS290LC was teamed with an 8-tonne JCB 8085 ZTC to remove the old track ballast and place it within easy reach for the JS290LC to then load one of WH Malcolm's colourful 30-tonne Terex TA300 ADTs. Both

The 65-tonne IHI Construction Machinery Limited CCH700 model with its 27m jib can handle all the lifting duties during bridge construction. Concrete is being poured to form one of the wing walls.

the ballast and soils from this area were to be used as fill material once the bridge reached its final resting position. The Terex ADT had a short distance to travel over a temporary bridge to reach the tip site, where a 2007 Komatsu D61PX dozer fitted with a VPAT blade was making short work of spreading the material. By 12.30pm, the first breach in the embankment was made and this allowed the Komatsu D61PX dozer to move to the south side to add some additional grunt at the large tip site, as the pace of the muck-shifting had picked up there.

While the excavators were busy on the cut, the SPMT team from world famous heavy lifting specialists Mammoet were starting to unload and couple up each of the modular trailers. Such was the global demand for its own 4,000 axle line-strong fleet of SPMTs, that Mammoet hired in some kit from another company. At the back of each SPMT is a massive 500hp M/Benz V8 twin turbo diesel engine driving two hydraulic pumps inside a large 7.2-tonne bolt-on power pack. This

power unit is capable of driving 40 axles and is under no pressure in this application as the longest trailers will have 18 axle lines in total. The SPMT trailers can reach a maximum deck height of 1.8m and each axle station has a powerful hydraulic 'suspension' ram with a 600mm stroke. However, on this job only 300mm of lift capacity is required. Each of the axle lines (four tyres per axle line, 272 tyres in total on this job) is rated at 40 tonnes. That said, Mammoet has set its own safe working limit at 36 tonnes and the total combined lifting force of the four SPMTs – with 68 axles – is 2,450 tonnes, leaving them with 650 tonnes of spare capacity!

BAM asked Mammoet to supply two 18-axle and two 16-axle SPMT combinations to lift and transport the 1,800-tonne bridge 8m to the east side and 30m north from the build area to its final position. Before the SPMTs could lift the bridge, the heavy duty abutment walls were braced with four large hydraulic bracing systems, each weighing 5 tonnes, and fixed to the inner walls so they did not bend or twist during the lift.

At 7am, and with the track clear, extra machines are deployed to excavate the embankment.

Pecking away with hydraulic breakers to remove the last metre of rock rising to 2.5m thick (left to right).

The Mammoet team starts to drive the SPMTs into position under the bridge, and support beams are fitted on the wing walls.

A 3:1 dry mix foundation on top of the rock bed. Note the strata rising on the far batter and 1–2.5m (L–R) steel plates being laid using magnetic attachments.

Now in its final position, a Liebherr 924 long-reach is used to backfill the inside of the wing walls

The excavators worked non-stop during Saturday, Sunday, and into Monday morning as the going got a bit tougher when they reached the bedrock – one metre thicker than expected in places. At this point the JCB JS290LC was using a large ripper tooth, and then switched to a hydraulic breaker. The Doosan DX300LC, Zaxis 280 LC and JCB 8085 ZTC were all engaged in digging and loading the broken rock into the waiting ADTs.

Once the team reached the desired level, a 3:1 dry mix was laid on top of the rock as a foundation for the bridge footings. Meanwhile, the small 8-tonne machines, with powerful magnetic attachments, were working with the Panoramic telehandler to lay steel plates over the uneven rock surface to make the drive of the SPMTs as smooth as possible. Once this work was completed – at 4.45pm on Monday – Mammoet's team leader started up all four power packs using just one hand control unit, and with 2,000hp at his fingertips, the bridge was lifted 300mm and started to slowly move forward from its home for the last 12 weeks, to its final destination. The move went perfectly to plan, including just squeezing the

Just six days since the start of the embankment removal the original track, sleepers and new blast are instated. This just leaves a few jobs to be completed, such as pre-tensioning the lines and welding the joints.

new bridge under the communications support frame. The operation only took 45 minutes to cover the 30m distance, which left Frank, Lydia, and the rest of the civil engineers to sign off on the exact resting position of the bridge.

Frank commented: "I've worked with the Mammoet team before and they've always carried out a very professional job, from planning through to the lift and carry phase of the operation; which went exactly to plan. They are a great crew to work with."

Over the next few days Babcock Rail worked quickly to reinstate the original track, and sleepers and new blast material was unloaded using special bottom dump railway waggons. By 25 September, this just left a few jobs to be completed, such as pre-tensioning the railway lines and welding the joints.

After weeks and months of meticulous planning and construction, Frank Young and his team were delighted – and somewhat relieved – that everything had gone to plan and they could hand back possession of the rail line on time, on budget and in readiness for an additional carriageway on the A723 to be constructed.

CHAPTER 28

M8 M73 M74 Motorway Improvements Project • Lanarkshire • May 2015

The £500 million, M8 M73 M74 Motorway Improvements Project is one of Scotland's largest transport infrastructure projects, which will include the construction of 12km of new M8 motorway and a new A8 all-purpose road incorporating upgraded sections of the existing A8. This will provide uninterrupted motorway and cut 18 minutes off peak journey times between Glasgow and Edinburgh. The project also includes a major redesign of Raith Junction, one of the busiest strategic connecting junctions within central Scotland's motorway network. The construction of an underpass below the existing Raith roundabout will create a free flow link on the A725, while three new bridges will carry traffic directly on to the M74 motorway. In addition, 30km of motorway carriageway is being widened and 16km of network upgrades will extend several existing segments of the M8, M73 and M74.

It is estimated that these improvements will affect more than 100,000 drivers, who will see an improvement in peak journey times across the three areas (M8, M73, M74) of the project. Other forms of transport have been accommodated, as the scheme will also provide 16km of new pedestrian and cycle paths.

This large project is another example of the significant infrastructure investment made in Scotland's transport network by the Scottish Government in the last few years, which includes the £1.4 billion Forth Replacement Crossing covered in Chapter 26 and the £320 million M80 Stepps to Haggs project, covering 18km of motorway, which was completed in August 2011 and is now shaving 15 minutes off journey times. The investment also includes the new rail link between Airdrie and Bathgate (opened in December 2010) and the M74 Completion project (opened June 2011) that provides an alternative route to the Kingston Bridge – in the heart of Glasgow – one of the busiest road bridges in Europe, carrying around 150,000 vehicles every day.

Scottish Roads Partnership (SRP), a Ferrovial-led consortium of Amey, Cintra, Meridiam, Aberdeen Asset Management, Ferrovial Agroman, and Lagan Construction Group, was chosen as preferred bidder for the design, build, financing and operation of the M8 M73 M74 Motorway Improvements Project. Alfredo Sobrino, construction manager for this project, is my host for the day, and our site visit starts just across from the SRP's main office at the A8 Chapelhall junction. At this location I find 12 of the 24 converted JCB 814 machines, which are drilling more than 3,000 test holes in a 3.5m box grid. They are expected to pump more than 80,000 cu.m of grout to eradicate old mine shafts and underground workings. This work has been subcontracted to Forkers Ltd, who carried out similar work during March–July 2013 at the Shawhead Junction (A8/A725) Coatbridge, where they were working on an advanced groundworks contract to consolidate worked coal seams.

The drilling and grouting work is scheduled to take four months, and Forkers used a large batching plant to mix a 9:1 blend of ash – from coal-fired power stations – water and cement, to form a low 1N/sq mm strength grout, and is pumping this mix at a rate of up to 400 tonnes per day. The ground here is being stabilised for the construction of a roundabout and junction to access the new M8. Given that the JCB 814 Super dates back to the late 1980s, the site supervisor said that they are simple and reliable workhorses, with good availability of spare parts, and a good platform for the drill head, where the operator stands and controls the drilling operation. The only time the operator sits in the cab is to track the

Some piling is carried using an 81-tonne Soilmec SR 70 hydraulic drilling rig.

machine to a new location and set the boom height and drill head angle.

My next stop is Raith Junction, where we meet structural civil engineer Marta Santos from the Ferrovial–Agroman–Lagan joint venture. Marta is responsible for piling operations at this section; to build an underpass by forming a 1,060m long wall made up of 1,600 interlocking male (reinforced) and female (non-reinforced) concrete piles, which will be finished with an outer cladding. To further support the wall, 500 post-tension concrete ground anchors will be constructed.

Piling is carried out using an 81-tonne Soilmec SR

Piling work at Raith, and a Kobelco crawler crane unloading reinforcing cages.

70 hydraulic drilling rig, operated by Cementation Skanska. This SR 70 has a 20m long auger and uses the Continuous Flight Auger (CFA) process, which is virtually vibration free and said to be one of the quietest forms of piling. This method enables piles to be formed in water-bearing strata, without the need for casing or bentonite. It is suitable for constructing piles in most types of strata, such as clays, gravels, sands, silts and soft rocks, which are the conditions found here on the former flood plain of the River Clyde.

CFA piles are formed by drilling to the required depth using a hollow stem continuous flight auger. After reaching the designed depth, a high slump (wet) concrete is then pumped in a continuous stream through the auger's hollow stem via a static concrete mixer unit, which is constantly topped up by a fleet of HGV mixer trunks fitted with the same size 8 cu.m barrel. While the concrete is being pumped, the auger is withdrawn at a controlled rate, removing the soil and forming a shaft of fluid concrete extending to ground level.

Cementation Skanska has also deployed a 73-tonne Kobelco crawler crane to unload and stockpile the reinforcing cages and then used to insert the cages into the hole filled with the fluid concrete. Reinforcing cages with lengths up to 18m and 900mm in diameter are installed with the assistance of a cage vibrator. Given the soft ground conditions, this technique is fast and cost-effective, as 12 to 15 piles per day is being achieved on a regular basis.

This site also requires larger diameter piles to carry the new bridges. These will be 1 to 2m in diameter by up to 30m deep and are being drilled by a 100-tonne Liebherr LB28 drilling rig, which is set up to use the casing method combined with bentonite as support fluid and, due to the large diameter, depth, and complexity involved, it can manage between two and three piles per day.

Leaving Raith, I head north on this massive construction project to see a new steel railway bridge being constructed next to the existing 'Cutty Sark'

Cut and fill operation at Eurocentral junction. Note the JCB 804s top right carrying out grouting and the D275, ADTs and Volvo grader keeping the haul road in good order.

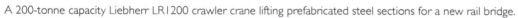
A 200-tonne capacity Liebherr LR1200 crawler crane lifting prefabricated steel sections for a new rail bridge.

bridge, as it is affectionately known after the whisky advertisement that adorned it for many years. Prior to my visit, three large Bauer BG40 piling rigs had been working to construct 2m diameter piles for the new bridge foundations. This new structure is being delivered in 45-tonne steel prefabricated sections from Severfield's Bolton factory. They are then lifted into position for bolting and welding by a 200-tonne capacity Liebherr LR1200 crawler crane. Once fully assembled, the 1,800-tonne bridge will be driven into position by the same method and heavy lifting company using a self-propelled modular transporter, as covered in Chapter 27.

A short distance away, on section 2, there is a major earthworks in full swing, where a pair of 70-tonne Cat 365L ME, (mass excavator configuration) prime movers are loading a fleet of articulated dump trucks (ADTs) to create a new six-lane (three lanes in each direction) motorway section of the new M8, which will run parallel to the existing A8 dual carriageway.

This section of the project, between Baillieston and Eurocentral, is being built 'off-line' to ensure there is no added burden or restrictions on the A8, and Alfredo and his team have more than 140 major and minor services (utilities) that will be affected by the project. Alfredo comments: "When working on a project of this scale, engagement and support from stakeholders, local councils, communities, contractors and suppliers become a key factor for the successful delivery of the project. Although disruption to road users is expected during the construction, we work very hard to minimise it as much as possible. Once construction has been completed the project will bring major benefits to road users in the central belt and the surrounding local communities."

A new A8 all-purpose road (APR) will be constructed alongside the M8, to accommodate the likes of agricultural tractors working the adjacent land and other vehicles that are banned from motorway use.

At the time of my visit, the big Cat 365 was sitting in a textbook position for loading on a 3m high clay bench. Loading trucks from a bench is more productive and fuel efficient as the return cycle is reduced, while the boom down function on this machine does not require any pump flow. As excavators go, the 411hp Cat 365 is king of the hill on this site and would not look out of place as the prime mover in a large quarry or even an opencast coal mine, removing overburden in a tight pocket of the site.

The shorter dipper arm on the ME spec 365 produces greater breakout forces, essential when removing large slabs of rocks encountered in some parts of this cut. With its 4 cu.m extreme bucket, I observed it loading the 40-tonne capacity Cat 740 ADTs in six quick passes, and five when loading the 35-tonne capacity trucks. This project is resourced with more than 200 earthworks machines, dozer, graders, rollers and 32-tonne capacity eight-wheel tipper trucks, which are moving some of the cut and fill material by road. They have a combined capacity to move in excess of 40,000 cu.m of material in a single day, weather permitting!

There is 3.5 million cu.m cut and 2.5 million cu.m of fill material to be moved across the site, and the remaining 1 million cu.m coming out of the cut will be used to build environmental bunds to screen the motorway and cut traffic noise to surrounding property.

This project is also looking to bring long-term economic benefits to Scotland; it is claimed that for every pound spent on this type of infrastructure project, it delivers £0.90p back into the economy, and there are also tangible benefits during its construction. One of Scotland's largest plant hire companies, for example, is operating the latest, state of the art and low emission Bell B30E ADTs on site, with another 13 recently purchased. At its height, there will be approximately 1,000 people employed through the project, with around 700 posts being filled by local contractors.

Graeme Reid, project sponsor for Transport Scotland, said: "As Scotland's busiest motorway, the M8 and the Central Scotland Motorway network is a main artery of the Scottish economy and the Scottish Government is committed to improving connectivity between our two biggest cities, cutting congestion, attracting inward investment and stimulating continued growth of our business communities. The project also offers exciting career opportunities for the scores of graduates and apprentices who are involved in its construction, and by investing in the skills and talents of young people, they are quite literally building a more prosperous future for themselves, their families and for Scotland."

Also supporting the earthworks are a large number of Komatsu D65PX, D85PX and Cat D6T dozers, and

A D6T fitted with a VPAT blade was working on a 12m high and steep 1:2 embankment to produce a finished surface with the 3D guidance system.

most of the Cat dozers are fitted with the Trimble 3D machine guidance systems. Alfredo explained that this guidance system has saved a great deal of time and money as the dozers are getting to the road foundation level quicker and are working to a high degree of accuracy. At the time of my visit, a new D6T fitted with a VPAT blade was working on 12m high and steep 1:2 embankment to produce a finished surface with the 3D system, eliminating the traditional setting out method of working.

It is not very often that you find three top of the range 27-tonne, 277hp Volvo G990 motor graders on a roads job. These large machines have been deployed

across the site to help keep the haul roads in tip-top condition and ensure the 50+ ADTs are operating safely and efficiently, as well as providing operator comfort and maximum tyre life. However, one of the subcontractors – Excav8 – has brought in a big Komatsu D275 AX dozer for when the ground conditions get a bit more demanding,

The D275 AX is Komatsu's answer to the famous Cat D9 dozer, tipping the scales at 51.5 tonnes and producing 450hp from its six-cylinder 15.2-litre engine. With its 4,440mm wide Semi U blade, the D275 was supporting the motor graders by dealing with the rougher parts of the boulder clay haul road and is also used for stockpiling material from the constant stream of ADTs as they move material from the cut area of Chapelhall to the fill area of Eurocentral junction.

This Komatsu D275 was sourced from the USA, where it had previously been used on a logging operation, hence the huge winch fitted to the back of the machine. With 12,500 hours on the clock, it had undergone a recent overhaul prior to arriving on site. The D275's operator said he was pleased with its performance, which is good praise, given that he normally sits at the controls of the bigger 71-tonne Komatsu D375A back in Ireland.

Over at the Shawhead junction, they were carrying out soil stabilisation, as the ground conditions were too wet for construction of a new M8 and Shawhead junction. Hard at work was a New Holland T6070 tractor coupled to a bonding agent spreader trailer that was discharging quicklime. A John Deere 8360 tractor was working alongside the spreader unit, and was powering a Stehr SBF 24-6 soil stabilisation machine; this unit is designed for dust-free milling of the soil and quicklime. During windy conditions or in an area where the lime dust must be kept to a minimum, the SBF 24 can be deployed on its own and has a 6 cu.m bonding agent tank so it can be a one stop shop.

Once the project is completed, it will provide long-term employment for more than 60 local people to maintain the new network for 30 years.

Soil stabilisation, as the ground conditions are too wet for the construction of the new M8 and Shawhead junction.

Forestry

When visiting such a diverse selection of applications, I find every day is a 'school day' as I learn so much about how each of the companies operate and what their sector contributes to the local or national economy. This is particularly evident when you read Banks Mining's contribution to the local economy. In the forestry sector, I was amazed to discover that more than 60% of the timber for paper, biomass and other timber-related products are harvested here in Scotland and – if you'll excuse the pun – activity is growing at a very healthy rate. And apparently trees grow faster in Scotland, due to the mild, wet climate!

Forestry activity contributes nearly £1 billion to the Scottish economy and employs more than 25,000 people when forest-related tourism is taken into account.

This section covers forestry operations at a port, loading and transportation in the forest, mountain biking trail building and a small JCB mini digger and its highly skilled owner operator working in the local community.

I head north to Inverness to see two timber handling machines from MultiDocker, who adapts Caterpillar excavators into some very impressive looking machines with massively tall undercarriage to load timber products into a fleet of ships.

Over on the west coast I visit a family-owned transport company that had a requirement for a dedicated timber loader to load its 44 tonnes trucks quickly and efficiently. With nothing available 'off the shelf' it built its own high performance hybrid machine, with great support from both the truck and excavator manufacturers and its longstanding engineering supplier.

It's not often you see a 1.6-tonne JCB mini digger fitted with a grapple on the end of the dipper arm, but that's what Wee Jim's Landscapes uses to feed his wood processor and harvest trees for his stockpile in Royal Deeside.

And, last but not least, I visit another highly skilled owner operator that has an international reputation for building mountain bike trails around Scotland and the rest of the world.

CHAPTER 29

Scotlog • MultiDocker CH85C • Inverness Harbour • February 2013

Scotlog is a family-owned and operated business dating back to the mid-1970s. It started as a trucking business transporting timber logs and then diversified into trading in timber. It started chartering ships to export the wood before growing the business to such an extent during the 1990s and 2000s that it was able to buy and commission new ships for its expanding fleet, trading under the name of Scotline, with seven boats ranging from 2,200–3,200 tonnes capacity.

Its shipping business operates from Inverness Harbour and it also has a road haulage operation in Inverness and Glasgow trading as Highland Haulage. Scotlog has made continuous investment in the most modern equipment for handling all types of general cargo.

MultiDocker's head office is in Norrköping, Sweden, and its cargo handling machines date back to 1981. Originally set-up as subsidiary of a large shipping business, it set out to create a world class cargo handling machine to suit its business needs and so MultiDocker was born. In building such machines the company chose to partner with Caterpillar in 1998 and gain the benefit of its products and RD resources. MultiDocker claims it uses more than 95% of Caterpillar's original equipment manufacturer technology in each of its machines. In 2008 it cemented its relationship further by partnering with Caterpillar's European dealer network, such as Finning UK, with the objective to sell every MultiDocker with a service agreement from the local Cat dealer. The machines are also a part of the Caterpillar certified rebuild programme.

Scotlog has a few firsts to its credit as far as MultiDocker is concerned; in January 2005 it took delivery of the very first MultiDocker unit in the UK. The machine was a 115-tonne class CH65B series II model and with five years of good reliable service, Scotlog traded up during 2010 to purchase the next generation of MultiDocker, a CH65C model that provided Scotlog with a new design and enhanced benefits.

The CH65C series came with an upgraded cab with more glass area to provide even better visibility than before. The C model also had a new hydraulic control system and joysticks with smoother more precise driver control, revised gantry and upper structure platforms with improved safety gates, ensuring even better access and egress. The gantry design came with an integrated 3,000-litre fuel tank. Scotlog's other first with MultiDocker came during January 2012 with the purchase of a top of the range 130-tonne CH85C model and the very first machine off the production line.

Stuart Catto, Scotlog's director of operations, explains the approach taken with MultiDocker: "In late 2004 our port activities were expanding to a point where we were spending about £160,000 a year hiring rope-operated cranes to load and unload our ships, so we headed to Sweden for a few days to see both MultiDocker and competitor material handlers in action. The advantages were obvious, as the hydraulic machines are much more productive and require less staff to operate; and having reviewed the market for a cargo handler, we decided to purchase a MultiDocker machine as it's based on a standard Caterpillar excavator, with all the associated benefits of Cat design, reliability and service back-up, and MultiDocker put a very compelling deal on the table. I can count on one hand the number of times the two MultiDockers have encountered any down time worth talking about, and as far as the CH85C machine is concerned, it's been extremely reliable since its arrival."

At the time of my visit, the monster CH85C, with 950 hours on the clock, was due a second service at 1,000 hours. Stuart has a fixed price service agreement

Produced locally, a 3,000 cu.m shipment of oriented strand board is being loaded into *Scot Venture*, bound for Varberg, Sweden.

with Finning UK and rates the aftercare provided by Finning's Muir of Ord depot highly. He said: "The local Finning depot is only 12 miles from the port and while we have very little unplanned down time, when we do need a quick response, the Finning engineers deliver day or night."

The big CH85C is a highly versatile machine with a number of different attachments at the operator's disposal, and was using a modified 4 cu.m clamshell bucket to load locally produced spruce and pine woodchips into the hold of the *Scot Isles* – rated at 3,200 tonnes capacity – which was bound for Sweden's papermaking industries.

This is a busy port, and the CH85C operator, from his lofty position, spotted the next inbound boat, *Scot Venture*, which was steaming under the Kessock Bridge. The massive CH85C, moving at 3.5kph, proceeded to travel the 300m to the other end of the dock to receive an inbound boat and switched attachments in order to load 3,000 cu.m of chipboard that was bound for Varberg, Sweden.

Meanwhile, back at the *Scot Isles*, with the hold doors now closed, the smaller CH65C swung into action to finish the loading job off using a SMAG HMG28 hydraulic grab, which is capable of lifting 4 to 5 tonnes of pulp logs on to the boat's deck in one pass; an easy lift for the 15.2-tonne rated capacity CH65C.

Stuart explained that since buying the MultiDocker units the number of staff needed to load a single boat has significantly reduced from 12 to just five, including the driver, and the boat can be turned around in just six hours compared to 14 hours using the previous rope cranes. MultiDocker claims a pulp log handling performance of 400 to 650 cu.m per hour for its top-line CH85C model.

Scotlog's long-standing MultiDocker operator, James Fraser, is very pleased with the MultiDockers and comments: "They have completely transformed the way

The extremely tall undercarriage allows HGVs to pass through or be loaded directly under the machine.

we load and unload boats. The visibility is great and has really improved the safety of the operation, as there are now only four dockers below to look out for. The telescopic cab and the front-mounted camera are really good to use when loading the material into the hold of the boat as you can see right into the work area. Being 10m off the ground takes a bit of getting used to at first, but both machines are so stable they soon put you at ease; even with an 8.4-tonne forklift at full stretch, you wouldn't even know it was lifting it!''

As would be expected, both MultiDockers come with overload protection systems, and their lifting capacity values have been calculated in compliance with SS-EN 130 01 regulations. In the case of an overload situation, the protection system will not allow the dipper arm stick to move outwards, therefore preventing an overload in the boom cylinders and uncontrolled lowering of the arm.

James also praised the comfortable cab, which is fitted with the usual refinements, such as a radio and CD player, air suspension seat, and air conditioning. The cabs are also equipped with Roll Over Protection Systems (ROPS) and Falling Objects Protection Systems (FOPS), such as the Margard safety glass to the front windscreen and right-hand side window that provides the operator with good protection against falling and penetrating objects. MultiDocker claims this safety feature is unique on the market.

The CH85C is required to lift the company's heavy duty forklift truck on a regular basis to unload chipboard positioned within the ships' holds, but the heaviest load the machine has handled is 15-tonne concrete slabs; again this is well within its rated lift capacity of 22.4 tonnes. James mentioned the electronic joysticks provide him with three options, with 21 variable settings. He normally uses the lower power, soft setting to deal

The CH65C starts loading the *Scot Isles* vessel with pulp logs using a SMAG HMG28 hydraulic grab.

The CH85C cab is fully extended 8m before loading a lift truck into the hold.

The CH85C cab has offset travel levers and glass cab floor for better visibility during loading operations.

The big CH85C had finished loading spruce and pine woodchips into the hold of the *Scot Isles* using a modified 4 cu.m clamshell bucket.

with very heavy loads, as this provides a greater degree of control and helps to ensure the load does not start to swing as it is lifted and lowered.

MultiDocker claims a 10% increase in hydraulic pump pressure over the standard Cat excavator machine and James confirmed both MultiDocker machines do not lack power or speed. The CH85C is equipped with Cat's C18 engine, which has mechanically actuated electronic fuel injection, and the 18-litre turbocharged engine produces a maximum claimed power of 530hp at 1800rpm and meets Stage IIIA exhaust emissions.

One of the main advantages of the CH85C over the smaller CH65C MultiDocker is the extra reach. Stuart explains: "Inverness harbour has a deep 5m berthing pocket, and with the tide and ships' draft – depending on its loaded state – to consider, the ship can move by as much as 5m in relation to the quayside. That means lifting an 8.4-tonne lift-truck was close to the limits of what the CH65C could handle in this application. So, on occasions we had to wait for the tide to drop before commencing loading operations. However, the CH85C has no such difficulties in this regard."

The two MultiDockers are versatile beasts, as both machines can handle a large variety of attachments such as clamshell buckets, log grabs, lumber spreaders, lifting frames and hook lifts. Each attachment is changed in a few minutes by releasing a single coupling pin and unplugging the hydraulic quick release hoses.

Both MultiDockers have massively tall undercarriages, which can accommodate an HGV and/or trailer being parked directly under the machine, which facilitates either loading or unloading with only 90 degrees of slew angle. The operator has a great view directly into the truck body, via the cab's large front screen, glass floor or the camera mounted under the cab. The camera is connected to a large colour monitor system, manufactured by Loke in Sweden. This also provides a clear view of the materials being handled.

For different applications, MultiDocker can supply three standard undercarriage types. The CH85C is fitted with 5m high MG8000C spec; this undercarriage is 6.5m wide (centre to centre), track pads are 1m wide, track frames are 7.57m long and the undercarriage weighs, including crawler tracks, an impressive 54.4 tonnes – that is about half the total weight of the machine! Due to the modular design, MultiDocker can respond to customer requirements to tailor the height and width of the undercarriage raiser and the length of the trackframes to suit the application. MultiDocker also uses some of the same parts that Caterpillar uses in the trackframes, such as the tried and tested travel motors.

The car body of the undercarriage also serves as a large fuel tank, 5,200-litre capacity in the case of the CH85C model and a 3,000 litre set-up in the case of the CH65C. However, the machines run off the standard Caterpillar fuel tanks – 1,240 litres on the CH85C. To refuel the machine's tank, the operator simply connects a fuel transfer pump hose to a quick coupling on the car body tank. Refilling the car body tank is carried out at ground level via another quick coupling. Stuart reports that, on average, the two big material handlers burn about 50 litres of diesel per hour.

Scotlog's CH85C is fitted with the shorter of the two stick and boom options, at 14m for the boom and 11m on the stick; long-reach options are 16,250mm boom and 12,800mm stick. Not surprisingly – due to the incredible 23.1m reach option – the CH85C is fitted with a 17,700kg counterweight in comparison to the 11,650kg one found on the standard Cat 385C excavator. And while MultiDocker uses many standard Caterpillar parts, the front end equipment is designed

and manufactured in-house, along with the gigantic undercarriage components.

The two machines operated by Scotlog are from MultiDocker's Port line of products, which are designed specifically for port operations and take port tides and vessel size into consideration as well as optimal visibility from the operator's cab. With that in mind, MultiDocker offers four options on operator cab mounting: the standard machine cab; a cab raiser – which sits the cab in high fixed position; cab forward, which puts the cab forward in a fixed position; or the fourth option – as in the case of both of Scotlog's machines – is an extendable cab, which on the CH85C can be hydraulically extended by an extra 4m, making an incredible 8m in total from the centre of the machine to give a clear view into the ship hold or dock area.

The other visually outstanding feature of these two Port Line MultiDockers are the undercarriage gantries and generously proportioned upper structure catwalks made from galvanised steel to withstand the harsh salt water operations. MultiDocker offers many options on its machines, such as fire extinguishers, autolube system, rear-view cameras and so on. Scotlog, with safety in mind, fitted an extra feature of its own – a lifebuoy ring!

Stuart explained the CH65C was delivered to the harbour in just three sections, however, the bigger CH85C came in eight parts – two tracks frames, upper structure, car body, gantry, boom, stick and counterweight – and it was a joint effort by MultiDocker and Finning engineers to assemble the machine. The Scotlog machines have been painted in the Scotline corporate blue colour scheme, rather than the standard MultiDocker/Caterpillar yellow.

During my visit there were two boats to be serviced, but not every day is that busy, in which case the CH65C is almost relegated to back-up machine status. However, Scotlog has about 190 boat movements per year, so the two machines can be used fully to provide a cost-effective fast turnaround of the ships and with a flat UK economy at the time of my visit Stuart was happy to report a 5% growth year-on-year for the shipping business. Clearly there is sufficient capacity in hand for the two MultiDockers to handle any future upturn in business activity.

CHAPTER 30

Peter McKerral & Co Ltd • Hybrid JCB JS220 • Argyll • February 2015

Peter McKerral & Co Ltd is a long-established haulage company based in Campbeltown, on the west coast of Scotland, and operates a large fleet of articulated heavy goods vehicles (HGV), hauling livestock, animal feeds, fish and timber products. A substantial proportion of the fleet is involved in transporting timber in the Argyll area, to ports such as Ardrishaig, Sandbank and Campbeltown.

Due to the vast quantities of timber used – for manufacturing and biomass purposes – McKerral has long favoured the use of a dedicated timber loader to load its 44-tonne trucks quickly and efficiently. That said, for self-loading applications, a small number of its timber trailers are equipped with a standard vehicle-mounted crane.

It is estimated that the UK will produce an average of 16.5 million cu.m of softwood per annum over the next 25 years. Scottish forests produce the largest proportion at 10.5million cu.m; with strong growth in the Scottish private sector expected to increase from 4.6 million cu.m to an estimated peak of 9.1 million cu.m by 2031. With business continuing to expand in this sector, McKerral required a highly productive machine to cope with this rate of production, so when its previous timber loader, a 2006 MAN 6×6 TGA chassis and Epsilon 250L 13.5m reach crane was due for renewal after eight years of good service, Donnie McKerral decided to upgrade to a bigger machine – a much bigger machine.

Donnie is a partner in the business and hands-on operator and manager; he had very specific requirements when looking for a replacement timber loader. It would be working from the company's home base in Campbeltown and cover an extensive area around Argyll. It had to be self-propelled, have superb off-road capability and load a steady stream of 44-tonne vehicles as quickly and as safely as possible, as the timber is often being loaded straight from the trucks into waiting boats heading for both domestic and international markets.

With nothing available from any original equipment manufacturer, Donnie decided to build his own machine. The search started with a high performance off-road vehicle and cab chassis, and after many meetings and about six months of discussions with his usual trucks suppliers without success, Donnie found a Mercedes–Benz dealer – Western Commercials, based in Glasgow – who could meet all his requirements in the form of a near military grade Arocs eight-wheel drive monster. This investment makes it the first Mercedes–Benz truck to enter the McKerral fleet.

Donnie's vision for this project was to mount a big excavator on to the truck chassis to achieve the desired loading performance. While looking for a vehicle, Donnie was also having parallel discussions with the leading excavator manufacturers to find a machine that would fit the bill. Again this was no straightforward exercise, as he only wanted to buy the upper structure and boom of a standard excavator. Donnie found Scot JCB, also based in Glasgow, to be very helpful throughout the search for the right machine. JCB was confident that its base machine could handle a massive industrial crane boom and dipper, so he decided to place an order for a UK-manufactured, 22-tonne class JS220. The JS220 was selected as the biggest machine that could achieve the hydraulic power for the bespoke crane boom and remain within the axle weights of the vehicle. It was ordered in material handling spec, as this model has a high-lift cab fitted at the Staffordshire factory.

With the truck and excavator now selected, Donnie then turned to his long-term supplier of timber bodies, Fergus Mitchell, to pull the project together. Fergus is the

owner of Forest & Field Engineering, based in Blairgowrie. They are the Scottish dealer for Hiab's range of Loglift Jonsered equipment, which covers forwarder/harvester and lorry loader cranes, as well as static industrial cranes. For this project it selected the largest static industrial spec'ed Jonsered J2990-160 crane in the range, with an original boom length of 8,250mm and a dipper of 5,300mm, plus hydraulic extension in the dipper arm of 2,550mm. The Jonsered crane boom was cut and grafted on to the JCB excavator boom. Fergus was also tasked with designing a sub-chassis for the excavator slew turret mounting and for making sure all the components would work in a tight operating envelope.

Fergus used a mix of computer-aided design software, particularly around the stabiliser legs and 2D scale drawings. However, when it came to the final sign-off meeting, with Fergus, Donnie, Scot JCB and Western commercials, Fergus decided to make a simple 2D scale model, so that everyone around the table could visualise the articulation of the dipper arm and boom in its working envelope.

Once the Arocs truck arrived at the workshop, Fergus quickly started to build a sub-frame from 300mm high steel box sections and welded the JCB slew turret into position, using some 350 high tensile bolts to secure the sub-frame to the vehicle chassis. Six weeks later, when the JS220 upper structure arrived on the back of a low-loader, Fergus had a mobile crane waiting to lift it straight off the trailer and on to the newly constructed turret and sub-frame of the Arocs.

During the design process Scot JCB recommended moving the top mountings of the excavator boom lift rams to a lower position as it had used this configuration on a demolition machine with great success. A new section was built below the neck of the boom; this allows it to achieve the maximum lift height for loading operations and also enables the log grapple to be parked close to the machine for transporting purposes.

Once the Jonsered boom was fitted, Scot JCB's team of engineers supported the project by reconfiguring the standard hydraulic control patterns to a pattern with which Donnie was more familiar. Hydraulic power for the four stabilised legs, eight hydraulic rams in total, (four to extend and four to lift) is from the redundant undercarriage travel motor circuit. The stabilisers are operated via the hammer line foot pedal (to activate

the flow) and a handheld remote control unit, which is commonly found in mobile crane applications for the same purpose.

Once operational, Scot JCB also provided on-site support to tweak the pressure settings to obtain the desired hydraulic speed, grab performance and power to meet Donnie's requirements. Fergus and his experienced team completed the project in just six months. The final painting of the company's colourful blue and red livery was carried out by Courtney & McMillan Ltd, based at Broxburn, and the JCB yellow of the JS220 make a smart looking combination.

Donnie said: "From the initial discussion through to the on-site support, Scot JCB and the factory have been very supportive about what we were trying to achieve and they did not shy away from getting heavily involved with such a bespoke order and aftersales support. Fergus and his team have done another cracking job of designing and building this machine to a very high standard."

Although the JCB JS220 is mounted on a HGV chassis, the design of this hybrid vehicle means that it cannot carry a load and has therefore been registered under special types vehicle rules, and with the DVLA as a mobile crane. This brings with it a number of advantages: a normal four-axle truck is limited to 32-tonne gross vehicle weight, but in this configuration, as a mobile crane, it can go to 41 tonnes, with the vehicle–excavator package finally weighing in at 37 tonnes.

At the time of my visit I met up with Donnie just off the A819 in the Ardteatle Forest near Loch Awe and joined him in the cab of the Arocs for a 3-mile drive up the forest roads to the loading point. He explained about its performance. "I'm really pleased with its performance. It has great on and off-road capability and has sufficient power from the six-cylinder 450hp engine. Fuel consumption is about 5 or 6 mpg, which isn't too bad given its weight and time spent off highway; it's what I expected."

With a 16-speed manual gearbox linked to the latest shift-by-wire system to provide more effective power-assisted gear shifting at the push of a button, the gearbox has a low ratio option, making 32 gears in total. With permanent eight-wheel drive, it is not about to get stuck off-road anytime soon. The Arocs was ordered with the lowest possible cab height – to accommodate the boom parked over the cab roof – and Donnie has chosen a

This Arocs has permanent eight-wheel drive, making it the ideal vehicle for this application.

sleeper cab, not that he spends that many overnights in the cab, but on rare occasions he likes to have that option available.

On reaching the loading area, Donnie applies the handbrake, switches off the engine and proceeds to climb steps fitted on the nearside of the truck to access the JS220 cab. Everything is now powered from the 4.8-litre, 179hp JCB EcoMAX T4i/Stage IIIB engine, which JCB claims uses up to 10% less fuel than its previous Tier 3 unit, partly due to the fact that the EcoMAX engine produces high torque at a low 1500–1600rpm.

With his foot on the hammer line pedal, the 4.25m-wide stabiliser legs are quickly deployed, via the remote hand controller. However, Donnie needs to be mindful of the o/s/f leg (which is set further back due to the position of the Euro 6 exhaust) to make sure it clears the swing radius of the JCB rear counterbalance. With all four jacks planted firmly on the ground, he can unpack the boom, dipper and log grab. Before loading commences, Donnie activates the high-lift cab, which

rises it into the air by a further 2.1m. The base of the cab is now about 4.5m off the ground, which makes the operator's eye line well over 5m. This helps to provide a great view on to the timber stockpile and the decks of the waiting semi-trailers.

As you might expect, the JCB cab is made in-house and to high standard, with all the creature comforts, such as climate control, stereo, air suspension seat, etc. The view out of the large two-piece right-hand window is excellent and it is also fitted with a rear-view camera, which helps Donnie to guide his drivers back to the rear bump-stops. Donnie can operate the loader and manage the operation from his two-storey high 'office'; and since not many log loader cabs could offer this level of comfort during a 12-hour shift, it was an important aspect of the selection process. Donnie uses a hands-free Bluetooth phone and where there is no mobile phone reception he can contact his truck drivers via a CB radio to help ensure there is always a truck lined up ready to be loaded.

This log-loader has a 16.5m reach.

Modified front screen grille and on-board weighing calculates each pass and total loaded, 27.5 tonnes per truck

The JS220 is positioned on the chassis some 3m back from the rear of the Arocs and was loading back to back. With the remaining 13.5m of reach the operator could easily load the front deck section of the 13.6m long semi-trailers.

The machine was ordered with proportional controls on the joysticks. These are now fitted as standard on the JS220 and are used to control the rotation and log grab functions. This JCB has a level 2 FOPS screen and roof guards, which has been modified by removing a portion of guard from the front screen area to provide an unrestricted view to the work area. Donnie believes that having an extra-long boom and dipper there limits the risk of logs coming anywhere near the cab, but he also felt it was wise not to remove it completely. It also serves to protect the cab from cosmetic damage, such as branches encountered along narrow roads and forest tracks.

Time is money in this business, as they are paid by the tonne, however Donnie does not get paid a single penny for anything over 44 tonnes, so there is no incentive to run over weight. With that in mind, there is a load cell coupling on the end of the dipper. This is connected to the in-cab HTP 2500 load indicator – specially developed for use on timber cranes – so Donnie can place exactly 27.5 tonnes of timber on to the trailer bed.

The JS220 is positioned on the chassis some 3m back from the rear of the Arocs, and was loading 6.2m long logs, weighing nearly 2 tonnes, with a large 0.8 sq m Loglift Jonsered grapple. When loading back-to-back, on the narrow forest roads, Donnie could load the front deck of the 13.6m long semi-trailers with ease. And with the front to rear stabilisers set at 5m apart, 4.25m wide and its all up weight of 37 tonnes, this loading tool is very stable in operation.

This new machine has a significant performance advantage over Donnie's previous loading tool, as he reckons that during a long shift, man and machine are capable of loading 1,500 tonnes of logs, which is an increase of 500 tonnes and is a good match for the boats with the same carrying capacity. I observed him load the first 27.5 tonnes in about eight minutes. Contrast that with an operator working in another part of the forest

The last of the 27.5 tonnes of logs being loaded, with Loch Awe in the backdrop.

A massive pusher frame is fitted at the front of the Arocs to assist drivers if they get stuck in the forest roads.

who took nearly 25 minutes to load his trailer using a standard vehicle-mounted crane with hydraulic power being supplied by the truck's power take-off unit.

In addition to the large bump stops fitted on the back of the chassis, Donnie has fitted a massive pusher frame to the front of the Arocs. This is to assist his drivers if they get stuck in the forest roads, and no sooner was this mentioned than the third truck of the day lost traction on a sharp rising incline a short distance away. When the driver radioed for help, Donnie quickly packed up the boom–dipper into its stowed position, fired up the big Arocs and with its eight-wheel drive system running on Michelin's deep treaded X Works tyres on the rear and XZYs on the front drive axles, he got the truck moving again with a quick push and no drama involved. This is an impressive piece of kit!

Donnie is delighted with its all-round performance and reckons that from his initial vision through to final completion, the whole project took two years. When you consider the large budgets and timescales some manufacturers have for research and development of new machines, what has been achieved here is remarkable.

Company Profile

The company was formed by Peter McKerral in 1936 and started as a farm business. He was joined by his son, Donnie (Snr), and his wife, Catherine, in 1968. In 1972 it branched into haulage, starting out as livestock, hay and straw merchants. In 1990 the harvesting of timber in the area made the firm extend the fleet to include timber haulage. Donnie and Catherine's four sons, Peter, Donnie (Jnr) Bobby and Colin, then joined the business. The company took over other businesses in Lochgilphead and Strachur, giving scope to extend its work. Peter McKerral & Co. Ltd operates a fleet of 50 lorries consisting of Scania, Volvo, M.A.N., Renault and DAF brands. It has 70 trailers, mainly flat timber trailers, some with timber cranes, three cattle floats and seven fish tanks.

CHAPTER 31

Wee Jim Landscapes • JCB 8018 CTS mini digger • Royal Deeside • November 2015

Jim Anderson is the owner and operator of Wee Jim Landscapes. Jim started his career in 2002, working for the local authority in the parks department and learning his trade carrying out landscaping and ground maintenance work.

In 2006 Jim decided it was time to start up his landscaping company, going full-time in 2007, and with a leap of faith bought a 1999 pre-owned Hitachi EX15 mini digger at the start of 2008. This machine was reliable and had provided good service, but Jim felt it was time to trade it in for a newer machine. At this time, a new salesman, Lee Taylor from Scot JCB, just happened to call round to introduce himself and a conversation soon started about buying a replacement mini digger.

Jim agreed to purchase a pre-owned JCB 8018 machine in nice condition. However, Jim had only driven it for four hours when one of the tracks snapped on a Sunday. Jim had no hesitation in calling Lee on a weekend and, to Jim's delight, Lee arranged for a new track to be fitted on Monday morning. Jim comments on this experience: "When I called Lee he could not be more helpful, and arranged for the parts to be delivered next day. I simply cannot fault the aftercare that Scot JCB provide."

Jim has nothing but praise for Scot JCB service; recalling another incident, where he accidently damaged the main lift ram on a Friday afternoon, at the start of the May bank holiday weekend. After a quick call, the local dealer ordered the replacement ram direct from the UK factory; allowing Jim to pick the part up in Aberdeen at 7am the next day. This quick service resulted in Jim keeping his customer happy, as he needed the JCB 8018 all weekend to get the job finished: "Because it was the May bank holiday weekend, the digger would have otherwise been out of action until at least Tuesday or Wednesday, which would have resulted in the job being completed late and lost income. You just can't beat that kind of service," said Jim.

It is often said that the salesman will sell the first machine and service department will sell the next one. With that in mind, when it was time to trade in his first JCB, Jim bought another one, an 8018 CTS, in June 2013. Again it was pre-owned; a 2011 model with just over 600 hours on the clock, and was supplied with a new set of tracks, a new dozer blade and a set of powerful work lamps to help maximise productivity of the machine during the dark winter months this far north.

Jim has a number of strands to his business, such as landscaping, agricultural, and snow clearing work. However, at the time of my visit, he was busy preparing a large order of firewood – from seasoned softwood – for one of his long-established customers. We met at his home village, where he operates out of an ex-sawmill in Glenmuick, Ballater. In addition to his JCB 8018, Jim has a 2002 115hp John Deere 6520 tractor, coupled to a new Palax KS 35 Ergo firewood processor.

The firewood processor comprises a hydraulically powered feed-in belt, an easy to operate 15in long chainsaw and hydraulic log splitter. The standard configuration of the processor comprises a 2.2m long and 20cm wide in-feed conveyor and a 4.3m long, swinging and folding discharge conveyor, as well as an automatic lubricator for the saw chain, with a maximum cutting diameter of 35cm.

Jim has two sources of timber to produce his firewood; he buys from local logging contractors, as his yard is large enough to accommodate a 44-tonne HGV, and he has an agreement with various local landowners to remove trees that have fallen as a result of high winds, which I will cover later on in this chapter.

The 115hp John Deere 6520 tractor is coupled to a Palax KS 35 Ergo firewood processor, which can cut and split 6 cu.m of wood per hour.

Jim loading the log rack, which has its own feed-in drive motor.

Jim has fitted the JCB 8018 with a grapple attachment, which was originally supplied with a short equaliser bar. This initial arrangement got in the way when working with a bucket attached, so a longer 12in bar was ordered as well as repositioning the anchor bracket further up the dipper arm. Using the grapple, Jim grabs roundwood from his stockpile and slews through 180 degrees to load timber on to his hydraulic feed-in log bed frame. Despite its 1.6-tonne light weight, the little JCB has a lift capacity of 250kg (at 1.5m high and with the blade lowered) that allows it to lift the large diameter, 3m long logs into the frame with ease.

Once the log bed is full, Jim switches off the 8018 and fires up the tractor to use its power take-off (PTO) to operate the wood processor, which quickly, cuts, splits and sends the firewood up the conveyer belt to load the dropside truck and or trailer at a rate of up 6 cu.m per hour. It has capacity leftover to fill a couple of bulk bags for stock. The work undertaken by the JCB 8018 in the timber yard is easy going compared to what man and machine has to overcome to get some of the wood to his yard in the first place!

The next part of our visit involved a short trip. Jim used the JCB's grapple to load a small logging trailer into the back of his pick-up truck and then drive the JCB 8018 on to a flatbed trailer and safely strap it down for a short trip to a forestry plantation on the Abergeldie estate. At the time of my visit, Jim was assisted by his part-time sawman, Norman Nicol, (at weekends) to clear some windblown timber. However, through the week, Jim will use the chainsaw himself to cut trees into suitable lengths for transportation.

After a ten-minute drive, we reached our destination and the equipment loading process started in reverse. Once unloaded, Jim hooked the logging trailer on to the towball that has been bolted to the 8018's dozer blade, before making our way into the woodland. Jim informed us that his usual source of timber, on the Abergeldie estate, has a decent access road for his digger to harvest the trees. However, the woodland we had just entered had no such luxury; it looked like the last machine in here had been a big forwarder machine that had caused some deep ruts in the ground along the route Jim was about to travel over. The only area of clearance had old tree stumps to overcome, as the rest of the area

was densely covered in mature trees that had to be manoeuvred around!

I must confess overcoming large ruts and tree stumps looked a big challenge for a little mini digger, as the low ground clearance is only 150mm, under the car body of the undercarriage. I feared that the pictures taken here had more chance of ending up in the picture post section of a magazine, i.e. bogged down or overturned! I could not have been more wrong, as Jim's skill behind the controls of the digger was simply amazing. In order to improve machine control in these difficult situations, Jim has fitted a foot pedal kit, costing just £70, to the travel lever arms to allow him to keep both hands firmly on the joysticks at all times, as he used all the front end equipment, track and slew motors, and dozer blade simultaneously to overcome many obstacles along a 200m distance to reach the fallen trees.

Once Jim got close to the first tree, he unhooked the trailer on a piece of level ground, then tracked in a bit further and used the grapple to lift and carry the sawn logs back to the trailer. During one cycle, Jim had to put the 8018 into one of the large ruts – which was full of water – to reach the timber. In order to get the digger out of this extreme position he simply used the log as a ground support, bracing the log across the deep rut to help lever the machine back on to more stable ground. He made it look effortless!

With a full load of eight logs on the trailer, Jim hooked up the trailer again and started to weave his way back to the main road. The trailer suspension has a large degree of articulation and wide high-flotation tyres to cope with the very challenging ground conditions; the suspension is very similar in design to that found on the rear axles of articulated dump trucks. Jim is very pleased with its performance and it is a good match for the mini digger and this type of application. Once Jim reached the main road he found a safe place to offload the timber, which allowed him to go back for a second run for the remaining wood. When finished, he loaded everything back on to the flatbed trailer and pick-up truck to head home. The newly stockpiled timber would be transported back to the yard using the front end loader on his big John Deere tractor to load on to an agricultural tipping trailer later on.

Jim explained that for some of his landscaping and ground maintenance work he might hire in a bigger midi

With no access road built, the JCB 8018 has numerous deep ruts and tree stumps with which to contend.

Jim takes deep ruts made by a heavy forwarder in his stride, and uses the log to provide a ground support to get the machine back on to level ground.

With the trailer uncoupled, Jim is ready to load the first 2.5 metre long logs – anything shorter than this and the wood will fall through the trailer side bolsters arms.

digger or a backhoe machine as and when required, and all of his tree work is carried out by NPTC and LPAF qualified arborists.

As part of the local community, Jim's other role in life is to help ensure the road network stays open during the winter months. He is contracted by the local authority to fit a massive directional snowplough to the front of his John Deere tractor to support snow clearing duties on the A939/B976 and also on the highest public road in Scotland, the A93 Spittal of Glenshee, at 2,198ft above sea level, to help keep the traffic flowing to the busy Glenshee Ski Centre from the Ballater area.

Given the soft ground conditions, the JCB 8018 is remarkably stable at full reach when lifting large 2.5m to 3m long logs.

When it comes to maintaining the JCB 8018, Jim will carry this out himself and explained that due to high track hours for his line of work, he will change the track motor gear oil more frequently than is normally recommended. When compared with other machines, Jim feels the cab of his JCB 8018 CTS machine offers him all the space and comfort he could wish for, and is also pleased with its fuel consumption as it uses less than 5 gallons of diesel per day. And with great aftercare service from the JCB dealer, Jim is already making plans to buy his next mini digger from JCB's compact products factory in Staffordshire and Scot JCB.

CHAPTER 32

Hitrak Ltd • Glenlivet Estate • Tomintoul • February 2013

Mark Hedderwick runs his family business, Hitrak Ltd, from Beauly, Inverness-shire where he offers plant hire and environmental contracting services across Scotland. Mark has significant experience in designing, constructing and maintaining footpaths, car parks, roads, mountain bike trails and other recreational forestry works, which are often in environmentally sensitive areas. The company is a regular contractor to the Forestry Commission Scotland, Crown Estates, Scottish Woodlands Ltd, various community forest trusts including Abriachan and Forres, National Trust for Scotland, SNH, Cairngorm National Park and Highland Council.

Crown Estates and Forestry Commission Scotland are helping to put mountain biking in Scotland on the world stage. They have been growing a network of fun and challenging trails to make Scotland one of the top places on the planet to visit and their goal is to offer world-class trails, stunning scenery and facilities. Their efforts are not without success; since 2002 Fort William has hosted an international cycling union (UCI) world cup event each year and amassed an impressive haul of awards during that time. These include the UCI Best World Cup award, which the Fort William event has won more than eight times.

Furthermore, Scotland was recognised by the International Mountain Biking Association (IMBA) in 2005 and 2006, which declared Scotland as a 'Global Superstar' due to its outstanding natural environment and availability of constructed and natural trails, coupled with some of the most forward thinking outdoor access legislation anywhere in the world. This resulted in some 400,000 visitors per year and generated more than £9 million for the local economy in the south of Scotland alone.

Hitrak Ltd has been building mountain bike trails since 2004, including parts of the prestigious Fort William course. However, Mark started off by building footpaths and car parks within a forestry setting, and in 2004 paid a visit to the ScotPlant show. In doing so he realised the potential for a tilt-rotator hitch for trail building, and purchased a demonstrator Volvo EC140B excavator fitted with an Engcon tilt-rotator. Mark reckons he is one of the first Engcon users in Scotland and at the time of my visit he was using the same tilt-rotator unit, albeit after some 3,000 hours, it had been repined and bushed and was attached to its second machine, a 15-tonne class zero-tail swing New Holland–Kobelco E135.

I met up with Mark and the rest of the crew at the Glenlivet Mountain bike project, which is a joint venture between David Rutherford of D&I Rutherford Ltd and Hitrak Ltd, who had previously worked closely on a number of similar projects. The contract was of a design and build nature, with the initial design and planning permission already granted, they subcontracted Paul Masson, of Cycletherapy, an experienced and renowned local trail designer, to carry out the micro design of the Glenlivet trails. With 19km of new trails, this was the first significant project in Scotland for some time. Other subcontractors were Chris Rogerson and a local forestry contractor, Graeme Collard.

Paul's brief was to create a mix of red and blue graded trails that would not only be challenging and fun to ride, but have a degree of safety built into the layout. Paul uses small red or blue flags to plot the path of the trails and is on site most days. However, having worked with Mark and his team for a number of years, they do not need to be micro-managed, because as Mark puts it: "We speak the same language." Mark mentioned that all his staff either ride mountain bikes or enduro motor bikes, so the team has a depth of trail building knowledge that is in constant demand.

Mark owns and operates a number of zero-tail swing machines that are also useful for working in confined spaces on house building or civil work and perform well in both applications.

I observed Mark forming a new trail using mostly type 1 – as dug – material, found just below the topsoil. He was working in among tightly packed trees, which necessitated pulling the boom and stick right in, close to the cab, to minimise the swing radius, and avoid damaging any trees or branches. Given this is time consuming and precise work, I asked Mark why he doesn't just cut the trees down? His response was: "That's the essence of good trail building," as once they have finished the trails should look natural and free flowing and are sometimes less than a metre wide in places to achieve this effect.

Mark is full of praise for his five-year-old New Holland–Kobelco E135, which is in stunning condition. He comments: "Most people would think about using

either 5- and 3-tonne excavators for working in tight confined spaces, however in my experience this 15-tonne excavator with the tilt-rotator can handle most jobs, and due to its size it's a highly productive machine; that's why we had three 14/15-tonne class and two 8-tonne class machines on this contract."

Mark's other excavator on site is an 8-tonne Hitachi 85-3 US zero-tail swing model and is quick to point out that there is only 300mm of difference in width between the two machines but nearly half the weight, with the E135 sitting at only 2.7m wide!

Mark is also pleased with the stability of his E135, as the majority of the trails have to be formed using a full-cut-bench-technique. That is where the dozer blade on the machine is essential to make sure the excavator is firmly planted on the steep slopes. As mentioned previously, Mark has been using an Engcon tilt-rotator for a long time and it has been ultra-reliable. He was quick to praise the Scottish dealer, JCC, for a responsive and professional service when needed.

The company has enjoyed a good experience with

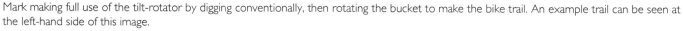

Mark making full use of the tilt-rotator by digging conventionally, then rotating the bucket to make the bike trail. An example trail can be seen at the left-hand side of this image.

this Japanese-built New Holland–Kobelco machine, having operated an original blue Kobelco 135 machine in the past, and Mark is delighted with its build quality and reliability. He made two specification changes to the E135: by fitting a dozer blade, a shorter dipper arm to keep the machine in balance – due to the extra weight of the Engcon unit – on this zero-tail swing machine and by removing the standard cab lights and replacing them with a bank of high intensity LED work lamps, which are a real benefit when working in the pitch black of the forest during the hours of darkness.

While Mark has no issues with his main dealers, his preference is to have all servicing and repairs carried out by Calum Fraser (ex Finning/Caterpillar), owner of Kilmorack Plant Services.

I first witnessed the benefits of tilt-rotators on a visit to the M80 extension, near Glasgow, and at the time they were mainly used to form batters around drainage ponds. In this application, they really use the maximum performance of the Engcon kit; Mark had his tilt-rotator attached to a 0.75 cu.m bucket to skilfully dig, spread, shape and compact the material to form the raised trails,

and is an impressive sight. To cap it all, Mark is not using any proportional controls to operate the tilt-rotator as he is happy with the standard hydraulic auxiliary line controls buttons on the New Holland joysticks and a foot-operated hydraulic pressure control valve. That said, he did spend £3,000 for the supply and fit of the Engcon proportional joysticks found on his Hitachi 85-3 US machine.

If Mark needs to increase breakout forces, for hard digging, then the Engcon hitch is fitted with two pick-up points which allows him to quickly drop the tilt-rotator off, and he reckons the hitch has only been used for about 3,000 hours during the 9,000 hours he put on each of the two machines he has previously owned.

He comments: "When building trails, the beauty of the Engcon hitch is you can strip out the top layers, then infill with as dug material, profile the track, narrow it in to about 600–700mm wide and landscape the edges. And not only that, if you used a 5-tonne machine it would need about four passes to move the same amount of material, which the E135 will move in a single pass, so

Gregor Macleod operating a New Holland E135B and Power Tilt attachment to load a Marooka MST800 tracked carrier at a nearby borrow pit.

Mark owns and operates a number of zero-tail swing machines fitted with tilt-rotators, which are useful for working in confined spaces in the woods.

the bigger machine is far more productive and will move a great deal of material in a short space of time as well as doing all the delicate work, too."

Once the trails have received the full tilt-rotator treatment, the rest of the team moves in to carefully shape and compact the material with either a Bomag double drum pedestrian roller or the Amman whacker plates at its disposal. While Mark recognises pedestrian rollers are not favoured now, he will not use ride on machines due to the nature of trail building – with steep angles to form – as the potential for a roll-over is too high.

When submitting their tender, Mark and others involved used their local knowledge of the ground conditions, and took into account the use of material (type 1) to be used along the trails and to make sure they would be on time and budget at the end of the project. With that in mind, they had Gregor Macleod operating a New Holland E135B model at a nearby old borrow pit that had been previously worked to maintain the forestry access roads to extract a small

amount of material when required. This machine is fitted with a Power Tilt attachment, which Gregor reckons is another useful bit of kit to use when constructing trails as it means there is no need to place material under the tracks to obtain the correct angle. And while it's not as all singing, all dancing as a tilt-rotator, it does a good job overall.

The stone quarried at the borrow pit was used as a top dressing on areas where the as dug had a higher clay content and was transported via a Japanese-made Marooka MST800 4,300kg capacity tracked carrier, which is capable of speeds of 7.5mph to cover a mile long distance between the pit and where Mark's team was working.

The old borrow pit location will also serve a secondary purpose; at Mark's suggestion the pit will form a new car park and a sunny café area for visitors to the new trails. With 19km of new mountain bike trails, this will be a welcome additional visitor attraction to the nearby towns of Tomintoul, Aviemore and Cairngorm national park to further boost forestry tourism.

An 8-tonne Hitachi 85-3 US zero-tail swing model forming bike trails with the use of an Engcon tilt-rotator.

Postscript

And finally …

I was given a piece of advice a number of years ago, "that the only constant in life is change"!

As covered in the foreword, this book represents a snapshot in time of companies I have visited and in some cases revisited, along with some very skilled, experienced and helpful people I've had the privilege to meet.

Here are just a few of the changes of which I'm aware:

You may have noticed in the quarries chapters that Lafarge changed its trading name to Lafarge Tarmac. This was under a merger by Anglo American UK operations and Lafarge UK operations, which came together during March 2013; and with Lafarge wishing to merge with Holcim in 2014, both companies eventually sold their shares to CRH Plc and it was rebranded Tarmac with the old blue circle logo returning in 2015.

The other big change was at CA Blackwell, shortly after my visit to Glensanda. CA Blackwell Group was bought by Hargreaves Services at the start of 2016, and if you recall from reading the mining chapters, Hargreaves also bought the assets of ATH Resources and Scottish Coal in 2013.

Just a few months after Volvo construction equipment had acquired Terex – truck products – and the Motherwell factory, this led to a rebranding to simply Terex Trucks, and added one more to the many trading names over the last 60 years.

Mark Hedderwick, of Hitrak, had a long-held ambition to own and operate a 'spider excavator' for some of the more extreme jobs in forests and mountains around Scotland and took ownership of a Kaiser S12 machine at the start of 2016, the first machine in the UK.

Now that the book is finished, I am looking forward to getting back on site again to capture more incredible machines and the people who operate them, and with that in mind, if you have an interesting machine and or application in Scotland, please get in touch with me through *Earthmovers Magazine*.